Nuclear 2.0

Nuclear 2.0

WHY A GREEN FUTURE NEEDS NUCLEAR POWER

MARK LYNAS

Published in paperback in 2013 by
UIT Cambridge Ltd
PO Box 145, Cambridge CB4 1GQ, England
www.uit.co.uk
+44 1223 302 041

First published as an Amazon Kindle Single in 2013

Design by Jayne Jones

Front cover reproduced by kind permission of Amazon.co.uk.
Front cover design by Richard Boxall.

ISBN: 9781906860233

Printed by CPI Group (UK) Ltd, Croydon CR0 4YY

Disclaimer: The advice herein is believed to be correct
at the time of printing, but the author and publisher
accept no liability for actions inspired by this book.

10 9 8 7 6 5 4 3 2 1

Contents

Introduction

Humanity is currently on course to double or even triple the carbon dioxide content of the Earth's atmosphere by the end of this century. Our current level of about 400 parts per million (ppm) of CO_2 is already higher than at any time during the evolutionary history of *Homo sapiens*. By 2100 the carbon content of the air could reach levels not seen for as long as 50 million years, pushing up global temperatures by 4-6°C (7.2-10.8°F) and transforming our planet beyond all recognition.

To try to visualize what this 'carbon bomb' might mean in reality, a few years ago I published a book called *Six Degrees: Our future on a hotter planet*. For analogues of the future, I was forced to examine the geological past – in particular an event over 55 million years ago called the 'Palaeocene–Eocene Thermal Maximum'. This 'thermal maximum' was seriously hot – rainforests grew up to the poles, Antarctica was green rather than white, and the Arctic Ocean was full of seaweed and as tepid as the Mediterranean. It's likely there was not a speck of ice anywhere on the entire planet: not on the poles; not even on the peaks of the highest mountains.

Would humanity have survived on such a planet? We may be about to find out. Emissions continue to rise each year as the developing world emerges from poverty, and vast new reserves of fossil fuels – most recently in the form of trillions of tons of shale gas and oil – continue to be discovered and squeezed out of the ground using ever-more-sophisticated drilling technologies. Despite rapid recent growth, wind and solar renewables still account for only about 1 per cent of global primary energy. If we are to regain any control over the climatic thermostat – and defuse the carbon bomb – we are going to have to change course dramatically.

Writing *Six Degrees* left me feeling both depressed and scared. I knew that the 'Small is Beautiful' world view shared by many of my environmentalist colleagues – who insisted that we must cut our personal energy use and supply all our needs with wind and solar power – was dangerously delusional when faced with a growing and increasingly affluent world population living in ever-expanding megacities. If this shift back to a simpler lifestyle was truly what was needed to forestall catastrophic global warming, then humanity simply had no chance. Like my green-minded friends, therefore, I feared deeply for my children's future. At a professional level we all tried to maintain a cheerful façade, but I knew from private conversations that most other environmentalists also felt profoundly pessimistic.

It came as a shock to me to discover, while attending an energy conference hosted by Oxford University academics back in 2005, that I had entirely overlooked the world's most promising source of low-carbon power. There I learned for the first time that nuclear power provided 5 per cent of global primary energy and 15 per cent of global electricity (2005 figures), and that this reduced carbon dioxide emissions by at least 2 billion tonnes* per year. I was stunned. As an environmentalist I had come of age within a movement that regarded anything 'nuclear' as irredeemably dangerous and evil, yet its potential to help tackle climate change was undeniable. Even though I knew that doing so would anger and antagonize my green colleagues, I started to research and write about the real risks and benefits of nuclear power.

As I did so, I began to discover that most of what I had originally believed about atomic energy was inaccurate. I had thought that nuclear waste was an insoluble problem, that using civilian reactors raised the risks of nuclear war, and that radiation from accidents such as Chernobyl had killed tens of thousands or even millions of people. As I looked more closely at the scientific data, however, I found out that most, if not all, of the anti-nuclear ideas I had grown up with were either myths or misconceptions. In fact, here was a reliable energy source with virtually unlimited fuel, which could power entire countries while producing no CO_2 at all during its operation. I found myself in the difficult position of coming to believe that my children's future was threatened not just by the big fossil-fuel companies but also by

* Imperial equivalents are given for metric values throughout this book, but these are not included for tonnes, as the standard imperial ton is almost the same as the metric tonne. (In contrast, the 'short ton' used in the US is approximately 0.9 × 1 metric tonne.)

professional anti-nuclear Green groups, many of which were staffed by my friends.

This is not an easy subject to write or talk about. Anti-nuclear opinions are passionately and sincerely held, as mine once were, and should not be casually dismissed. Therefore, this book aims to explain, using the very latest factual data, why maintaining an anti-nuclear ideology is both ill-conceived and fundamentally incompatible with resolving the climate-change crisis. I present the argument using numbers rather than just assertions, because this is a numbers game: our task is to generate tens of thousands of additional terawatt-hours per year of power using energy sources which do not commit the planet to catastrophic global warming or otherwise threaten the biosphere. This is all about scale, and the scale of the challenge is increasing with every day that passes. All of us, environmentalists included, must therefore move away from business-as-usual attitudes.

My conclusion in this book is ultimately an optimistic one. This is not another lamentation of doom. With an Apollo Program scale-up of nuclear and other low-carbon power sources we still – just about – have time to avoid the worst of global warming. But this will require a lot of burying of hatchets. Most importantly, if we are to confront the vested interests that threaten to keep this planet on its current trajectory towards disaster, the pro-renewables and pro-nuclear tribes will have to join forces. I urge anyone who is serious about global warming to read on with an open mind.

CHAPTER 1

How we got to where we are

Carbon emissions from fossil-fuel burning (ignoring agriculture, deforestation and cement production), which already total about 32 billion tonnes of CO_2 per year, on average rose by 676 million tonnes annually from 2000 to 2010. This is equivalent to almost twice the emissions of Brazil added to the global total during each twelve-month period.[1] Although this massive increase in fossil-fuel burning has caused disastrous pollution and health impacts in China and elsewhere, the dramatic rise in global CO_2 is actually a good thing in one respect, because it shows that developing countries are expanding their energy consumption in order to extricate themselves from centuries of famine and misery. In other words, we are losing the war on carbon precisely because we are winning the war on poverty.

The rise of the rest

Energy use, and in particular electricity use, is tightly correlated with human welfare. Access to modern energy sources – electricity and natural gas rather than charcoal, firewood or dung – saves hundreds of thousands of lives per year through avoided respiratory disease alone. Development is also the best insulator against climate extremes, which is why African droughts can kill tens of thousands of people from starvation and famine, while Australian droughts kill none. It is ironic but true that for developing countries currently the best way to protect against the future effects of climate change is to burn more fossil fuels and thereby accelerate their economic development, despite the obvious wide-scale negative consequences from smogs, traffic pollution, acid rain and so on.

The common simplistic answer to this conundrum – to dramatically reduce energy use in rich countries in order to compensate for growth in the developing world – simply won't work: the coming growth in energy demand is an order of magnitude more than could ever be saved even by the most extreme energy efficiency measures adopted in the West. Consider that the 19.5 million inhabitants of New York State consume the same amount of electricity as the 791 million people in sub-Saharan Africa:[2] satisfying this unfulfilled African demand up to American standards would require a 40-fold increase in the production of energy. In total, 1.4 billion people still do not have access to electricity today. All of them want it, of course, and will soon get it as their countries develop.

With the exception of a few war-torn states, a scattering of autocracies and theocratic outliers such as North Korea and Iran, every developing country in the world is today enjoying sustained growth. Africa, once the world's basket case, is now the world's fastest-growing continent – witness its 'lion economies', where per-capita income rose by 30 per cent in the last decade alone. (The Asian 'tigers' have helped; *The Economist* reports that over the past decade African trade with China has risen from $11 billion to $166 billion.[3]) This increased wealth has contributed to dramatically improved human welfare: secondary school enrolment is up by 50 per cent, while infant mortality rates have plummeted in almost every country.

Although Africa is in the early stages of the energy catch-up game, Asia has been powering ahead for much longer. Between 2001 and 2011, Indonesia increased its electricity generation by 79 per cent; Bangladesh by 150 per cent and Vietnam by 261 per cent. This coming to wealth and power – in both senses of the latter word – of the developing world is the great tectonic geopolitical shift of our time, dubbed by CNN's Fareed Zakaria as 'the rise of the rest'. As he writes: "The tallest building in the world is now in Dubai. The world's richest man is Mexican, and its largest publicly traded corporation is Chinese." Poverty, as defined by the number of people living on less than $1 a day, has plummeted by 40 per cent since the 1980s. Between 1990 and 2010, the size of the world economy tripled, with over half the growth coming in so-called 'emerging markets'.[4]

The epicentre for this developing-world economic expansion has of course been China. As Zakaria writes: "China has grown over 9 per cent a year for almost 30 years, the fastest rate for a major economy in recorded history. In that same period, it has moved around 400 million people out of poverty, the largest reduction that has taken place anywhere, anytime." He concludes: "The average Chinese person's income has increased twentyfold... China has compressed the West's 200 years of industrialization into 30."[5]

Coal reality

The vast majority of the new energy for this economic growth has been provided by coal. In China, half of all rail-freight capacity is currently used to transport coal from mines to power-plant furnaces. China's electricity generation more than tripled between 2001 and 2011 (an increase of 217 per cent), and, apart from small contributions from wind, nuclear and hydropower, all this growth has been driven by coal. As the International Energy Agency (IEA) wrote in the title of a recent slide presentation: "China is coal. Coal is China."[6]

China now uses about half the world's coal, and produces a quarter of global CO_2.[7] (The US produces just over a sixth of global CO_2, though it is important in equity terms to remember that US per-capita emissions of 17 tonnes still dwarf China's per-capita 5 tonnes.) China's consumption of coal more than doubled during the last decade, rising by 155 per cent from 2001 to 2011. The oft-quoted 'factoid' that China adds two coal-fired power stations every week to its grid is not quite accurate; the true figure for the last decade was 1.15 new coal-fired power stations per week. The total coal-generation-capacity addition between 2000 and 2010 was 450 gigawatts[8] – about 1.3 times the entire coal-power capacity of the US.[9] China adds the equivalent of the whole UK coal fleet every six months.[10]

China is coal, but coal is not just China. Coal is India too. As the 2012 IEA *Coal Market Report* put it: "Endowed with large coal reserves, a population of more than 1 billion, electricity shortages and the largest pocket of energy poverty in the world, India makes the perfect cocktail to boost coal consumption."[11] India is soon expected to pass the US as the world's second-largest coal consumer after China.

For coal, the only way is up. The IEA's 2012 coal report predicts: "The world will burn around 1.2 billion more tonnes of coal per year by 2017 compared with today. That's more than the current annual coal consumption of the United States and Russia combined." In terms of our energy mix, therefore, and despite tablet computers, synthetic biology and Twitter, we are really only midway through the Industrial Revolution. There is still a long way to go in terms of the rather old-fashioned human habit of extracting carbon-containing materials from underground and setting fire to them in various ingenious ways to run our machines.

Fossil fantasies

It is worth mentioning at this stage that there is no prospect whatsoever of us running out of coal – or indeed any other fossil fuel – in time to stop runaway climate change. China has the third-largest coal reserves in the world, after the US and Russia; India the fifth-largest.[12] At current rates of production, global coal supplies will last for 112 years, and no doubt vast new reserves still remain to be found.

Perhaps that is just as well, in human welfare terms. Although climate impacts are unlikely to drive resource wars any time soon, true energy shortages might well have done. Had China's wave of new growth begun to hit a true wall of 'peak oil' or 'peak coal', the geopolitical implications hardly bear thinking about. One might imagine the nuclear-armed militaries of the two world superpowers clashing over diminishing supplies of the fossil energy. It is therefore lucky for us all that the often-apocalyptic predictions of those who preached the 'peak oil' mantra have so far turned out to be wrong.

Instead of imminent fossil fuel scarcity, in fact, the world now faces an age of astonishing abundance thanks to the new technology of hydraulic fracturing, or 'fracking'. Having driven natural gas prices down to unprecedentedly low levels in the US over the last couple of years, the shale gas revolution now promises to repeat the same feat in areas such as China and Europe. Shale gas has provided undeniable short-term climate benefits by driving out the much more carbon-intensive coal from the US power sector, but shale is a double-edged sword. In the longer term, a major expansion in cheap hydrocarbons will surely make it harder to deploy more costly low-carbon alternatives and will thereby increase humanity's total carbon emissions.

Meanwhile, in the US the new talk is of 'Saudi America', as shale oil exploitation becomes the new frontier: by March 2013 the US had actually overtaken Saudi Arabia in total oil production, drilling 11.8 million barrels per day as opposed to 10.85 million.[13] "North America has set off a supply shock that is sending ripples throughout the world," IEA Executive Director Maria van der Hoeven said at the launch of a report on oil markets in May 2013. "The technology that unlocked the bonanza in places like North Dakota can and will be applied elsewhere, potentially leading to a broad reassessment of reserves."[14] In other words, the world is awash with new oil and gas supplies, and probably will be for decades to come.

Industrial society's dependence on fossil fuels is often rhetorically condemned by politicians and campaigners as an 'addiction': one that every US president dutifully pledges (and then duly fails) to kick. During any recitation of the list of fossil fuels' environmental and social ills, it is conveniently forgotten that they remain the foundation of the prosperity we all now take for granted in rich countries. Eco-romantics may fantasize about the pre-fossil-fuels economy of the past, but the reality of pre-industrial agrarian society really was slavery, war, famine, disease and a short average lifespan for everyone except a very privileged aristocratic few.

For the thousand years before fossil fuels were harnessed, humans therefore saw almost no long-term economic or population growth. Malthusian checks and balances were all too real. However, at about the turn of the nineteenth century, something dramatic began to happen. Look at any graph of life expectancy, population, GDP and CO_2 emissions over the last millennium, and marvel at how all four curves suddenly shoot upwards in about 1800, especially at the epicentre of the Industrial Revolution in the West.[15] This is not just correlation, it is causation. Fossil fuels allowed humanity to transcend previously hard-wired ecological limits that were the result of being tied to a precarious 'organic economy' with only solar-powered photosynthesis for life support.

CHAPTER 2

The carbon challenge

Fossil fuels may have liberated us from a crude agrarian existence, but now we are entering a different era. The global-warming crisis is sufficiently urgent that fossil fuels must be phased out and replaced with alternatives that can maintain an energy-intensive and growing human civilization without destroying the life-supporting capacity of the biosphere.

Solar and wind

The standard prescription for tackling climate change is to mobilize a combination of energy efficiency and solar and wind power. While efficiency is clearly a good idea in both theory and practice – you get more services per unit of energy – expecting this combination to actually reduce overall energy use is a different matter. Historically, greater efficiency tends to accompany an increase in overall energy use: Western economies today are generally twice as efficient as they were 40 years ago, but use far more energy in total. There are good economic reasons for this: efficiency reduces the price of energy compared with other factors of production, thereby stimulating demand for energy.[1]

That leaves wind and solar, in the orthodox view. Both are unambiguously low-carbon; I have no truck with sceptics on that point. Research I conducted with the environmental writer Chris Goodall, drawing on several months' data from the UK electricity grid and published in *The Guardian*, indicated that current wind power generation in Britain successfully replaces gas-fired generation (each hour of wind power is an hour of reduced gas burning) on a megawatt-hour for megawatt-hour basis, and therefore certainly does reduce overall CO_2 emissions. Somewhat to my surprise, our research almost exactly confirmed the aggregate CO_2 mitigation totals claimed by the wind industry.[2]

It is a myth that at low proportions of total power in the grid – 10 per cent for the UK, US and most other countries – wind intermittency results in fossil-fuelled power plants inefficiently ramping up and down. The additional intermittency is easily absorbed by current system overcapacity, and is less of a problem for grid managers than people switching on their kettles or electric coffee-makers during unscheduled television advertising breaks in major sports tournaments, for example. Yes, this might change with wind power at over 20 to 30 per cent of the total in the grid, but there is enormous debate about what the implications are and how they might be managed. For the foreseeable future, in my view, the more renewables we can add to the grid everywhere, the better.

There are other clear benefits from renewables, too. Unlike thermal power generation using the steam cycle, wind and solar photovoltaics use little or no water. (Some water may be needed for cleaning panels in the case of the latter when deployed in windy, dusty areas.) This makes these power sources especially appropriate for arid areas, where the strongest solar radiation is likely to be found. Renewables are also pollution-free in their operation, whereas burning coal and oil emits cancer-causing particulates, acidic sulphates, mercury and other toxins into our air and water. Although there have been cases of damaging river pollution in Chinese solar PV manufacturing, these are no more inherent to the technology than in any other comparable industrial process. Toxic metals used in some solar cells, such as cadmium and tellurium, can be recycled indefinitely, as can most components of wind turbines, including the 'rare earth' elements.

Unfortunately, renewables are now encountering a rising tide of political opposition. Aesthetic concerns are of course subjective, but some people seem to have an implacable gut hatred of wind turbines in their backyards; if ever the term 'nimby' were appropriate, it would be for these objectors. Likewise, while there clearly are serious issues regarding bird and bat kills in some specific areas of conservation concern, this will not – and should not – preclude the vast majority of both onshore and offshore wind-farm developments.

Some habitat-loss concerns with regard to specific species, such as the desert tortoise, have also affected the solar industry, but the truth is that the Earth has vast areas of barely inhabited deserts, and devoting a few tens of thousands of square kilometres to solar farms is not going to lead to a major crisis in biodiversity. Moreover, I cannot think of a single environmental objection to putting photovoltaics on rooftops, and there are extensive paved and urban areas worldwide that could be covered in this way.

Renewables revolution

As most readers will doubtless have heard, solar and wind power generation has expanded enormously in recent years. Between 2011 and 2012, wind-energy generation saw a worldwide growth of 18 per cent, and solar enjoyed an even bigger increase of nearly 60 per cent. This is the 'renewables revolution' in action – total wind-generated electricity grew by 200 per cent in the last five years, while solar grew by 1,200 per cent.[3] The costs of solar photovoltaic in particular have fallen dramatically, enabling a much more rapid upscaling of solar than once seemed realistic. In many countries, solar PV is on the verge of achieving the long-sought-after goal of 'grid parity', promising even more rapid expansion in future without the need for subsidies.

These figures have unfortunately generated a lot of hype, however, and to understand them properly we need to consider not just relative growth rates but absolute energy generation totals. Solar's meteoric 1200 per cent growth over the last five years took it from producing 0.01 per cent of global primary energy to 0.17 per cent – from the infinitesimal to the still rather small, in other words. Wind, with its 200 per cent growth, went from providing 0.3 per cent to 0.95 per cent of global primary energy. These are the numbers that matter in climate terms, because with such small amounts of energy production, current wind and solar power must necessarily be having only a small effect on emissions reduction. According to wind industry figures, wind farms saved about 350 million tonnes of CO_2 in 2011 – about 1 per cent of global energy-related emissions.[4]

Moreover, wind and solar are fighting to increase their relatively small shares of an energy-consumption pie that is growing rapidly larger each year. As we saw in the last chapter, the past decade saw massive energy-use increases in developing countries, and the vast majority of this growth came from coal. In the year 2011-12 (these figures were released in June 2013), coal added 101 mtoe, gas 73, oil 49, wind 18 and solar 8 mtoe to global energy supplies. ('Mtoe' stands for 'million tonnes of oil equivalent', an internationally accepted standard unit of energy.) Coal therefore added 5 times more to global primary energy than wind did last year, and 12 times more than solar. This was no annual blip: between 2007 and 2012, coal added 7 times more to primary energy than wind, and 30 times more than solar.

Let's double-check my numbers by looking at just electricity production – a reasonable thing to do, given that most coal, and all grid-scale wind and solar, are used to generate electrical power. In 2012, wind generated 2.3 per cent of global electricity; solar 0.4 per cent. World electrical generation rose by an annual average of 515TWh (terawatt-hours) between 2007 and 2012. (That's nearly a new Brazil [553 TWh/year] added to the global grid each year!) Wind accounted for about 14 per cent of this average annual increase, and solar for just over 3 per cent. Most of the rest was coal and gas. For just 2011-2012, in the midst of the renewables revolution, solar accounted for 8 per cent of the increased generation, and wind 18 per cent. The numbers are improving, but not by much, and certainly not by as much as the enthusiasts tend to claim.

Again, please do not read this analysis as being in any way anti-renewables. I am not suggesting that because renewables are still being outpaced by coal, this must always be the case – indeed it cannot be if we are to get a grip on carbon emissions. But I think this reality check about our current energy situation is essential to get the true picture of the sheer scale of the challenge ahead. As I said before, I'm a supporter of renewables; my intent here is simply to demonstrate with real-world numbers that it stretches credulity to argue, as most Greens still do, that solar and wind *on their own* can supply enough power to a rapidly growing civilization to solve climate change on the diminishing timescales we have available.

Energetic denialism

Here's Ralph Nader, back in 1998: "You want me to give you a book that shows how realistic solar is? You've got wind power, you've got biomass, you've got photovoltaic cells, you've got tidal, you've got all kinds of technologies now moving towards commercial viability." Or take Barry Commoner, even further back, in 1975: "All of the solar energy technologies that can replace gaseous, liquid, and solid fossil fuels are in hand; some are already economically competitive with conventional sources, and many are rapidly approaching that point." Sound eerily familiar? Yet this was during Gerald Ford's US presidency, even before Jimmy Carter heralded the new dawn by wearing an energy-efficient cardigan and putting solar panels on the White House.

Then there's Amory Lovins, who first popularized the term 'soft energies' for small-scale wind and solar in a hugely influential 1976 essay for *Foreign Affairs* called 'Energy Strategy: The Road Not Taken?'. Lovins was impressed by the 'Buddhist economics' teachings of E. F. Schumacher, whose own 1973 book *Small is Beautiful* inspired much of the early environmental movement. Schumacher proposed that the 'right' kinds of technologies were small, decentralized and controllable by ordinary people – an idea perhaps based on his own rather ahistorical notions of an idealized agrarian past. Lovins extrapolated this romantic misconception to energy technologies, and proposed that all US energy supplies might come from 'soft' sources within a few decades.

They didn't, of course. Lovins' 1976 projections were wrong by a factor of 60: he proposed that 'soft energies' might comprise 33 per cent of US primary energy by 2000; the true figure was 0.5 per cent.[5] And if Lovins' utopianism had little to offer back in 1976, it is surely utterly irrelevant today. A world population of 4 billion has grown to well over 7 billion, while global GDP has increased by two-and-a-half times. Though pessimists in the 1960s and 1970s expected famine to wipe out hundreds of millions due to overpopulation, in fact economic development and poverty reduction have accelerated. And almost all of the increased energy required for this came from the centralized, large-scale 'hard energy' technologies that Lovins wanted the world to reject.

Despite its long-term failure, Lovins' and Schumacher's ideal of humans using less energy is still central to environmentalist thinking. In a 1977 interview, Lovins said: "If you ask me, it'd be little short of disastrous for us to discover a source of clean, cheap, abundant energy because of what we would do with it. We ought to be looking for energy sources that are adequate for our needs, but that won't give us the excesses of concentrated energy with which we could do mischief to the Earth or to each other."[6]

This energy-reduction agenda was then and still is the exact opposite of what the world really needs. As Bill Gates has pointed out, the colliding imperatives of rapid economic growth and global warming make 'clean, cheap, abundant energy' the world's most desperately sought-after commodity. And the truth is that for the next decade at least, only a minor – though, one hopes, rapidly rising – proportion of this energy can plausibly come from wind and solar.

Like most environmentalists, I do not support a major increase in 'renewable' power from the other two large-scale sources, hydropower and biomass. Big dams destroy fragile riverine ecosystems, while biomass sources have included clear-cut swamp-land forests in North America[7] and palm oil from ex-rainforest land in Indonesia.[8] Both, then, have ecological impacts that drastically limit their additional scalability or desirability. Other than these, there is really only one other proven large-scale, low-carbon option currently on the table that can help us meet climate and energy targets along with wind and solar. And that is nuclear power.

CHAPTER 3

The N-word

For the majority of my career as an environmental writer and campaigner, I either ignored or disparaged nuclear power. My first climate change book, *High Tide: News from a warming world*, published in 2004, didn't mention the N-word at all, even as I ended the narrative about global warming impacts on places as far afield as Peru, Tuvalu and Alaska by imploring readers to "take personal action to reduce emissions" and "keep repeating the climate change message". My second climate book, *Six Degrees: Our future on a hotter planet*, did mention nuclear, but only in one sentence, most of which warned about "deadly accidents" and "the still-unsolved question of what to do with highly radioactive wastes" – all standard environmentalist talking points. The *National Geographic* TV adaptation of *Six Degrees* featured a whole sequence on nuclear – but this was all about the distant dream of nuclear fusion, not the current reality of nuclear fission.

Challenging the taboo

Still, during that Oxford University energy conference I mentioned in the Introduction to this book, a light bulb had gone on somewhere in my head, and, in my capacity then as a fortnightly columnist for the British magazine the *New Statesman*, I ventured a short article wondering whether nuclear ought perhaps to be reconsidered as part of the future energy mix in light of climate change. In the UK, I later discovered, the 25 per cent of our electricity supply provided by nuclear was being steadily reduced as ageing nuclear plants were decommissioned, threatening to push up carbon emissions. Should we not at least consider replacing them?

Fearful of the possible reaction, I hastened to reassure readers that "I'm not suggesting that nuclear is a panacea", and acknowledged once again the potential for "Chernobyl-style accidents or terrorist attacks", as well as the supposed "legacy of toxic waste for millennia". Moreover, I wrote, having more nuclear power should not compromise investment in renewables, and "can reduce carbon emissions only as part of a combined dash for renewables and energy efficiency, buying us time while truly clean energy systems are developed".[1] This was not exactly a full-scale assault on the core philosophy of the environmental movement.

Even so, within a couple of hours of my *New Statesman* piece being published online, hurt and angry responses began to flood in from friends and readers alike. Some questioned my motivations: I was now a fraud; a sell-out; an industry shill. Others were simple one-liners – "But what about the nuclear waste?" –

clearly intended to shut down debate rather than engage in any real discussion. One response I have never forgotten came from an activist friend who blamed me for having undermined her whole life's work in that one rather cautious 500-word essay.

Nuclear and environmental impact

Despite all the high emotion that nuclear power seems to cause, few people remember the rather prosaic fact that all a nuclear reactor does is generate heat. This heat boils water into steam, which expands to drive turbines, just as in any other thermal power plant. Unlike in a coal or gas plant, however, nuclear does not release CO_2 because it operates via fission rather than combustion of fuel. (There are emissions produced in the mining and refining of uranium, and via the concrete and steel of a power plant, but most experts agree that these emissions are comparable to those of wind power generation.) The problem is that the splitting apart of atoms of uranium to generate this heat releases highly radioactive 'fission product' elements, which need to be safeguarded in order to prevent them harming people.

I recently visited one of the UK's fleet of 'Advanced Gas-Cooled Reactors' at Hinkley Point B in Somerset, and was conducted on a tour by the plant's owner, EDF Energy. I was able to walk around right on top of the reactor core, and could feel a gentle humming as the gas circulated below my feet to conduct away heat being produced by the fissioning of uranium in the fuel rods. The core is so heavily protected by concrete shielding that I needed no special protection (other than the standard-issue hazmat suit and goggles that are mandatory for everyone inside the building) and the dosimeter my guide was carrying remained obstinately at a zero reading the entire time. Looking round the turbine hall afterwards, I could see a digital display indicating that the plant was generating 500 megawatts of power, enough to run a small city.

Nuclear power's singular environmental advantage can be summed up in the term 'energy density'. Consider that a golf-ball-sized lump of uranium, weighing just 780 grams (2 pounds), can deliver enough energy to cover *all* your lifetime use, including electricity, car driving, jet flights, food and manufactured goods – a total of 6.4 million kWh. To get the same energy output from coal would require 3,200 tonnes of black rock, a mass equivalent to 800 adult elephants and resulting in more than 11,000 tonnes of carbon dioxide.[2] The volume of this pile of coal would be 4,000 cubic metres (141,000 cubic feet): you can imagine it as a cube about 16 metres (52 feet) in height, depth and width; about the size of a large five-storey building.

The uranium fuel cycle is not the only way that nuclear power can be generated. Recently the potential of thorium as a nuclear fuel has generated a lot of interest – thorium is much more abundant than uranium in the Earth's crust, and could conceivably power advanced human civilization for tens of thousands of years. (All these heavy elements were originally fused in a supernova – an epochal explosion of a previous star more than 5 billion years ago, before our own sun and solar system came into existence. What we are doing in a nuclear power plant is reversing the energy cycle generated by that supernova.) Either way, nuclear power is the only means by which we can generate prodigious amounts of energy with only a tiny carbon footprint on the planetary biosphere.

The anti-nuclear movement

As can be seen from the outline above, a dispassionate examination of nuclear power yields no *a priori* reasons why environmentalists should be against it; quite the opposite in fact. (Concerns about nuclear accidents and radiation are dealt with in the next chapter.) Indeed, until the early 1970s many established Green groups were cautiously in favour of this burgeoning clean energy source: the Sierra Club, for example, embraced nuclear power in California as a better alternative to flooding scenic valleys for hydroelectric power. But something later turned this lukewarm support from the environmental movement into implacable multi-decadal hatred.

We could speculate endlessly about what this was. The historian Spencer Weart, in his magisterial work *The Rise of Nuclear Fear* (which I highly recommend to anyone interested in this area), suggests that opposition to nuclear reactors may have been a kind of psychological displacement effect, where suppressed fear of nuclear missiles found expression in activism against neighbourhood reactors. Certainly, many of the lifelong anti-nuclear-power activists, such as Helen Caldicott and Barry Commoner, started out as anti-nuclear-missiles activists – moving seamlessly from campaigning about radioactive fallout from nuclear weapons tests to trying to ban nuclear power. As Weart writes about Caldicott: "When she moved to the United States [in 1977] and found nobody there interested in bombs any longer, she began to fight reactors. Entire organizations took the same path."

The idea of the 'China Syndrome' – that a nuclear reactor faced with loss of cooling could somehow eject enough radioactive material to sterilize a huge area – was first promoted by the Union of Concerned Scientists (UCS), a Green group originally founded at East Coast universities, and helped win the UCS national recognition and a host more members. Many people wrongly thought (and still do think) that a nuclear reactor can explode like a nuclear bomb, and the idea of a 'nuclear explosion' causing similar levels of devastation whether from a reactor or a weapon was a psychologically powerful one, however implausible it might be in reality. The heart of this opposition came from nuclear fear, specifically fear of radiation, as an invisible, cancer-causing 'poison' which could harm – so it was claimed – millions of people through the operation of nuclear power.

Early environmentalists thought that radioactivity was somehow uniquely dangerous and polluting. In

1973, E. F. Schumacher wrote that radiation was "the most serious agent of pollution of the environment and the greatest threat to man's survival on Earth". As the Clamshell Alliance, formed to oppose the construction of the Seabrook nuclear station in New Hampshire, wrote in its 1976 founding declaration: "Nuclear power poses a mortal threat to people and the environment." A later 'Declaration of Nuclear Resistance' adopted in 1977 stated: "Our [anti-nuclear] stand is in defence of the health, safety, and general well-being of ourselves and of future generations of all life on this planet."

Against this background, the Sierra Club's initial refusal to campaign against California's Diablo Canyon nuclear plant could not possibly continue: in a top-level bust-up which nearly destroyed the Sierra Club, executive director David Brower resigned in 1969 and went on to found Friends of the Earth as a 'properly' anti-nuclear environmental group. By 1974 the Sierra Club had abandoned any attempt to carve an independent path: its board of directors duly fell into line with the new anti-nuclear orthodoxy, and the organization has held to it ever since.

One of the most widely read anti-nuclear books of the time was *The New Tyranny: How nuclear power enslaves us*, published by the Austrian journalist Robert Jungk in 1977. Jungk used his personal history as an anti-Nazi resistance campaigner during the Second World War to draw moralistic parallels between Nazism and nuclear power. He speculated that nuclear scientists secretly dreamed of creating an "improved race of human beings who can tolerate huge doses of radiation", and devoted a whole chapter entitled 'Citizens under guard' to the "ominous role nuclear power could play in converting a democratic nation into a totalitarian atomic state". Jungk later stood as presidential candidate for the Austrian Green Party, and helped to instil a loathing of nuclear power which is still widely held in both Austria and Germany today.

Conspiratorial thinking about nuclear power was also rampant in the US, where activists believed that power company executives would stop at nothing to contaminate and poison the population in the name of utility profits, as played out in fictional scenarios in Hollywood movies such as *The China Syndrome* and *Silkwood*. The extraordinary coincidence between the release of *The China Syndrome* and the accident at Three Mile Island (TMI) in 1979 led to a national panic: few in the media or general public were prepared to believe 'official' reassurances about safety and containment – even though they later turned out to be largely true. (Only a tiny release of radiation took place at TMI, far too small to cause any health effects in the surrounding population.)

On one occasion, opposition to nuclear power did spill over into outright violence: on the night of 18 January 1982, five RPG-7 rocket-propelled grenades (apparently sourced from Germany's murderous Red Army Faction) were fired across the Rhone river in

France at the unfinished containment dome of the Superphénix fast reactor, where earlier mass protests involving thousands had left dozens injured and one dead. The perpetrator, Chaïm Nissim – who remained anonymous for two decades – later became a Green Party MP in Switzerland, and today works as a member of a Swiss think tank promoting renewable energy. To date this episode is still the only terrorist attack to have been carried out against a civil nuclear installation in the world.

A world safe for coal

The success of the anti-nuclear movement in the 1970s guaranteed an increased use of coal for decades to come, as proposed nuclear plants across the Western world were cancelled and replaced by coal plants. There are countless stories with specific examples: one of my favourites is of the Austrian plant at Zwentendorf, a mid-sized nuclear power station. It was fully completed and then closed down in 1978 before it could generate a single watt, after anti-nuclear activists narrowly won a nationwide referendum. Today, although Austria has 60 per cent hydropower, it still burns coal and oil for a third of its electricity: had Zwentendorf and the other proposed nuclear plants been allowed to run by the nascent Greens, Austrians might have enjoyed carbon-neutral electricity for the past 35 years.

The Zwentendorf story has an irresistible coda: in 2009 it was 'converted' into a solar power plant. At the opening ceremony, backed by enormous Greenpeace banners declaring 'Energy Revolution – Climate Solution' and featuring Hollywood celebrities such as Andie MacDowell, 1,000 new solar photovoltaic panels were inaugurated, having been installed at a cost of 1.2 million euros. "From radioactive beams to sunbeams – a global symbol for environmentally friendly and sustainable energy for the requirements of the future," said the website.[3] A quick look at the numbers tells a different story, however: average output from the solar panels will be 20.5 kilowatts (enough to run 12 hairdryers, according to one wag[4]), whereas the 692 megawatts it would have generated as a nuclear station would have lit up Vienna.

One can chuckle at that kind of foolish hype, but less amusing is the history of Ireland's proposed Carnshore reactors, which were cancelled after protests, rallies and concerts were organized by anti-nuclear groups in the mid-1970s. A large coal plant was built instead, at Moneypoint in County Clare. Moneypoint's two chimneys, as well as being among Ireland's tallest constructions, are now the largest single-point source of CO_2 emissions in the entire country. Some of Ireland's electricity even comes from the only source worse than coal: peat. Peat-generated power is not only more CO_2-intensive than coal, but is based on the shameful industrial strip-mining of large areas of fragile and biologically irreplaceable raised peat bog.

In Spain, nearly 40 nuclear plants were proposed in the 1970s, but a strong anti-nuclear movement was successful in forcing a national moratorium in 1984 and only 10 were ever built. Spain today has 18 coal power plants, supplying a fifth of its power. In Australia, perhaps the most coal-dependent country in the world (despite its solar potential and abundance of uranium deposits), nuclear power is technically illegal, thanks to a thriving anti-nuclear lobby and a senate vote in 1998.[5] Australia's per-capita CO_2 emissions as a result are about 18 tonnes, even higher than in the United States, with coal supplying 85 per cent of domestic power.

In some places, half-built nuclear plants were converted directly to coal: one example being the William H. Zimmer plant in Ohio, whose containment building was converted to house a coal boiler instead of a reactor following protests and cost overruns in 1984. As the historian Spencer Weart writes: "Ever since the price of oil spiked in the late 1970s, wherever people refused to build more reactors almost every new electrical plant had been a coal burner." Each time this happened, determined anti-nuclear coalitions of thousands of environmentally concerned citizens melted away overnight once the embattled utility had agreed to change its proposed plant from nuclear to coal.

Allens Creek, Texas; Bellefonte, Alabama; Cherokee, South Carolina; Erie, Ohio; Hartsville, Tennessee; Satsop, Washington... The full list of cancelled US nuclear plants can be viewed on Wikipedia.[6] At Shoreham in Long Island a nuclear plant was fully built, as at Zwentendorf in Austria, and then was immediately shut down due to enormous public opposition, much of it paid for and fanned by the efforts of diesel-fuel delivery companies. Today it is a mausoleum – but had it been allowed to operate, it would have helped make New York a carbon-neutral city for the last three decades. I calculate the total capacity of all the cancelled nuclear plants to be about 140 gigawatts; roughly half the entire current installed coal capacity in the US.[7] More than 1,000 nuclear plants were originally proposed; had they all been built, the US would now be running an entirely carbon-free electricity system.

In the United States during the heyday of the anti-nuclear movement between 1972 and 1984, coal consumption by US utilities doubled, from 318 million to 602 million tonnes.[8] Although it is often claimed by Greens that their anti-nuclear activities were less important than the 1970s oil shocks and economic slowdown in forcing the cancellation of planned nuclear plants, during the period 1972 to 1984 the US added 170 gigawatts of fossil-fuelled capacity to its electricity grid[9] and consumed 74 per cent more coal-fired electricity[10] – hardly indicative of a major reversal in the growth of overall energy consumption. Certainly, the snowballing cost of nuclear plants was a major factor, but a significant proportion of those costs was being imposed by an ever-expanding nuclear regulatory burden, which slowed or stopped development of new plants and spent fuel repositories – even more than environmental activism did. Nevertheless, constant objection by vocal 'antis' generated increasing political risk and nuisance lawsuits, and thus caused years of delays.

That is not to say that the anti-nuclear activists liked coal. They said they wanted solar power, and the famous 'nuclear power no thanks' logo of course sported a smiling sun symbol. But just as they were spectacularly successful in stopping the growth of nuclear power, they were spectacularly unsuccessful in promoting the use of solar as an alternative. By 1984 the use of solar had risen from functionally zero to 0.002 per cent of US electricity generation.[11] The history of the anti-nuclear movement is therefore not lit by sunshine, but shrouded in coal smoke.

CHAPTER 4

The case against: nuclear accidents and radiation

Between them, the world's fleet of 400 or so reactors have run up tens of millions of pollution-free operating hours over the last quarter-century. Even so, no technology can be completely fail-safe under all circumstances. Think of dam failures, large bridge collapses, aeroplane crashes and chemical plant explosions: the best we can do is to minimize the dangers as far as technically possible through a determined and constant commitment to safety. In all these areas much higher safety standards now operate than was the case in the past, and nuclear power is no exception. Sceptics might ask whether the stakes are not higher in the case of nuclear: if things do go wrong, are the results not far worse than with other technologies? To try to address this, it is worth looking in detail at the two major civilian nuclear accidents that have released substantial quantities of radiation, at Chernobyl in 1986 and Fukushima in 2011.

Fukushima

The accident at Fukushima began as a consequence of the colossal magnitude-9.0 earthquake which shook eastern Japan at 2.46 p.m. on 11 March 2011. About 50 minutes after the earthquake, 14-metre (46-foot) tsunami waves swept over Fukushima's 5.7-metre (19-foot)-high breakwater before surging up the sides of the reactor turbine halls and other buildings. All the emergency diesel generators for Units 1-4, located in the basements of the turbine halls (ironically to protect them against earthquakes) had been flooded by seawater, as had the battery systems meant to back them up. Debris and mud covered the entire site. Stunned operators watched as one by one the station's lights blinked out, and warning alarms – also subject to a loss of power – faded away into silence. By 3.50 p.m. all instruments were dead, including the gauges needed to assess the water level inside the reactors themselves.

A well-understood design flaw in all 'light-water' reactors (those at Fukushima were GE-manufactured boiling water reactors, but the same criticism applies to normal pressurized water reactors) is that cooling them without external power is close to impossible – water must be pumped into cores which are normally kept at immense pressures in order to keep cooling water liquid at operating temperatures of several hundred degrees Celsius. The operators at Fukushima now had the unenviable task of somehow getting water into overheating reactor pressure vessels with no power, no instrumentation and no pumps, and within the context of a major nationwide disaster that had already killed nearly 20,000 people.

Although Fukushima's parent company TEPCO had immediately dispatched several mobile generator units from Tokyo, these had 250 kilometres (155 miles) to travel on earthquake-damaged roads, and quickly became snarled up in the traffic chaos of a massive natural disaster. Desperate for energy to restore power to the reactors' instrumentation, emergency crews scavenged batteries wherever they could find them – in the site's earthquake-proof emergency response centre, and later even from underneath the bonnets of the cars parked above tsunami flood level in Fukushima Daiichi's main car park.

By this stage, residents in a 3-kilometre (1.8-mile) zone around Fukushima Daiichi were being ordered to leave their homes and evacuate. The decision was the right one: although the operators did not yet know it, Unit 1 had already boiled dry, and both the temperature and pressure inside the reactor 'containment vessel' had shot up. With the entire core exposed and reaching temperatures well above 1,000°C (1,832°F), the sealed zirconium alloy 'cladding' used in nuclear fuel rods reacted with steam to form highly explosive hydrogen gas. This is another known flaw of light-water reactors in emergency situations, and to combat it the Fukushima designs were fitted with a venting system to release hydrogen and excess steam via a nearby chimney stack.

However, at exactly 3.36 p.m. on 12 March, before venting had been able to work properly, hydrogen that had been steadily accumulating on the top floors of Unit 1's outer building suddenly exploded.[1] The concrete roof and walls of the fifth floor were blown out, throwing chunks of masonry in all directions. Pieces of concrete slammed into a newly installed mobile generator and wrecked it, and other falling debris cut an electricity cable that was just about to energize the reactor's control systems. Five workers were injured by falling material but, remarkably, no one was killed. On TV screens around the world, news broadcasts switched to live feeds showing a large grey cloud engulfing much of the plant. As the evacuation was extended to a 20-kilometre (12-mile) radius, media headlines took on an increasingly apocalyptic tone.

Not surprisingly, many of the workers on site thought that they would probably die from radiation sickness as a result of their efforts, yet they courageously stayed at their posts. The plant manager, Masao Yoshida, at one point asked his colleagues to write their names on a whiteboard as a memorial in case they were killed.[2] In videos that emerged later, Yoshida is even shown raising the possibility that he and other older workers might need to mount a 'suicide mission', entering prohibitively radioactive areas themselves in a last-ditch bid to restore cooling to the stricken reactors.[3] These workers' selfless heroism is one of the great untold stories of the Fukushima crisis.

In reality, there would not have been much even a suicide mission could do. Analysis by the US Oak Ridge National Laboratory suggests that full-scale core melt in Unit 1 had likely begun to take place within as little as five hours from the onset of the emergency. Thereafter, molten core materials – about 140 tonnes of uranium oxide, zirconium and stainless steel – probably slumped down to the bottom of the reactor vessel.[4] Glowing red-hot like a lava flow, this molten mass then melted through the steel base of the reactor vessel and dropped on to the concrete floor underneath – a real-world China Syndrome meltdown scenario so feared by many. Yet the outcome was very different from that depicted in the movie: in reality the molten core materials penetrated about 65 centimetres (26 inches) into the 2.6-metre (8.5-foot)-thick concrete floor before they began to cool down and re-solidify.[5]

Within the next few days, both Units 2 and 3 at Fukushima Daiichi also suffered core damage, with perhaps a near-total meltdown also in Unit 3. At 11.01 a.m. on 14 March, preceded by a peculiar orange flash, an even bigger hydrogen explosion tore apart Unit 3's outer building, throwing radioactive material even higher into the air than the Unit 1 detonation. Hoses and fire trucks were damaged or put out of action by falling concrete, and the workers had to start from scratch once again. On 15 March, Unit 4 also blew up, at the same time as a still-unidentified noise took place somewhere inside Unit 2's containment, releasing the greatest amount of radiation of the whole event.

Fukushima health impacts

This all sounds pretty disastrous, but what gets lost in all the drama is just how extraordinarily effective were the measures taken to prevent damage to people's health. During those nightmarish few days, as Units 1 to 3 melted down in quick succession and massive explosions blew apart Units 1, 3 and 4, only six workers were exposed to a dose in excess of 250mSv (millisieverts) of radiation, which in both the US and Japan is the absolute maximum dose permitted for nuclear workers in life-saving emergency situations. In addition, 161 workers were exposed to a dose of between 100mSv and 250mSv,[6] while the average exposure for the several hundred other TEPCO workers was about 25mSv.[7]

What do these numbers mean? They are certainly far below levels that can cause radiation sickness, for which the threshold is generally around 2,000mSv. The United Nations Scientific Committee on the Effects of Atomic Radiation (UNSCEAR) discussed Fukushima in May 2012 and concluded: "To date, there have been no health effects attributed to radiation exposure observed among workers, the people with the highest radiation exposures. To date, no health effects attributable to radiation exposure have been observed among children or any other member of the population."[8] In other words, not only have there been no fatalities resulting from Fukushima, but there have been no radiation-related injuries or medically identifiable health effects either. This is a very different picture from the one most people got from the constant media coverage during and after the event.

In fact the only documented deaths attributable to Fukushima actually happened through panic-stricken attempts to avoid radiation. At least 60 hospitalized patients died during the evacuation, some because they were left behind in severely understaffed medical facilities when most nurses and doctors fled in terror from the looming 'radioactive cloud'.[9] Although one should not lightly attribute blame for emergency measures taken during the confusion of a crisis, it seems very likely that had these patients stayed in their beds with full medical attention, many of them would still be alive today. Their deaths can therefore be attributed to fear of radiation, rather than to radiation itself.

For the general public as a whole, however, experts agree that the evacuation was a necessary precaution, and certainly ensured that people living in the area around Fukushima Daiichi received much lower doses of radiation than they might otherwise have done. I'll quote at length here from an UNSCEAR press release issued on 31 May 2013:

> On the whole, the exposure of the Japanese population was low, or very low, leading to correspondingly low risks of health effects later in life. The actions taken to protect the public (evacuation and sheltering) significantly reduced the radiation exposures that would have other-

> wise been received, concluded the Committee. 'These measures reduced the potential exposure by up to a factor of 10. If that had not been the case, we might have seen the cancer rates rising and other health problems emerging over the next several decades,' said Wolfgang Weiss, Chair, UNSCEAR Report on Radiological Impact of the Fukushima-Daiichi accident.[10]

The World Health Organization (WHO) has also studied the Fukushima accident, and released a major report in February 2013 looking at potential future health risks. The 28 February 2013 press release stated: "For the general population inside and outside of Japan, the predicted risks are low and no observable increases in cancer rates above baseline rates are anticipated."[11] 'Low' is not the same as zero, however, and for particular population groups in the most affected town, the WHO did project small percentage increase risks in certain cancers. For all solid cancers it projected an increase of 4 per cent for females exposed as infants; for breast cancer an increase of 6 per cent for females exposed as infants; for leukaemia an increase of 7 per cent for males exposed as infants; and for thyroid cancer an increase of 70 per cent in females exposed as infants.

This latter finding made headlines around the world, but the newspapers mostly neglected to report the context: the lifetime risk for females of thyroid cancer is 0.75 per cent, so the projected 70 per cent increase would bring the total lifetime risk up to 1.25 per cent.

Similarly, the lifetime statistical solid cancer (i.e. not leukaemia) risk for a female infant might rise from 29 per cent to 30.1 per cent because of Fukushima. Strangely, the WHO conducted this analysis on hypothetically exposed persons in the affected area – it remains unclear how many children (if any) actually received the radiation doses on which the cancer risk projections are based, because most were quickly evacuated elsewhere. The Japanese government strongly criticized the WHO report on this basis, for raising unnecessary fears among an already traumatized population.

I visited the Fukushima area with the film director Robert Stone in May 2012 during filming for the movie *Pandora's Promise*. (This book was originally inspired as an accompaniment to that movie, in which I appear along with other pro-nuclear environmentalists.) We spoke to refugees living in temporary housing in nearby Minamisōma town, many of whom were restricting their children's hours of playtime outside because of fears of radiation. Yet the readings we took with our handheld dosimeter were not substantially different from the ones I get in England outside my own house in Oxford: 1.5mSv per year as opposed to 1.05mSv per year in Oxford. Readings in the evacuated areas were much higher, however: in Futaba town I found a ground-level hotspot which gave a reading of 2,130mSv/year just next to the hospital car park. Chest-level general readings were lower, in the order of 40-100mSv/year, but still double or triple the maximum dose below which displaced residents might be allowed to permanently return.

Radiation reality

Context is everything in this debate, and radiation numbers are often bandied around without any indication about what they mean. It is vital to understand that radiation also comes from the air and soil around us, from our food, from the sky, and from natural radionuclides such as potassium-40 in our own bodies. Together with medical radiation (from X-rays and scans), on average this adds up to 2-3mSv per year per person. The natural occurrence of potassium-40 has led to the concept of the 'banana-equivalent dose', because bananas are naturally high in potassium. A human adult on average contains 7,000 becquerels of ionizing radiation, meaning that there are 7,000 radioactive decays per second in our bodies: eating a banana temporarily adds about another 15Bq, about 0.0001mSv.

In various countries, health spas are sited around hot springs that are high in naturally radioactive radon, including those at Ramsar in Iran, where the highest levels of natural radiation in the world have been measured. On a web page about the issue, scientists can be seen taking readings as high as 1,250mSv/year from an inside wall of a house, while ground-level readings outside are about 113mSv/year,[12] and scientific reports suggest that 2,000 people in Ramsar accumulate doses of between 10mSv and 260mSv per year.[13] (Just 20mSv/year is enough to trigger an evacuation under international guidelines for nuclear power plant accidents.) A scene in *Pandora's Promise* shows a man with arthritis burying himself in dark sand on Guarapari beach in Brazil: the sand is high in naturally occurring radioactive thorium, and film-maker Robert Stone obtained a dosimeter reading equivalent to 342mSv/year in that location.[14]

I smile, therefore, when I hear anyone worry about trivial sources of radiation such as scanners at airports: in a recent opinion piece in the *New York Times*, for instance, Nathaniel Rich wrote: "I have never walked through an airport body scanner – or, as I think of it, 'the cancer machine'."[15] As the Health Physics Society states in a useful factsheet, actual scanner doses are in the order of 0.00001-0.00005mSv per scan. "This amount of exposure is well below any level of concern and, in fact, is less than 1 per cent of the radiation you receive from natural sources in a single day or less than two minutes of airplane flight."[16] If Mr Rich is so worried about radiation, in other words, he should avoid flying altogether, because a six-hour flight will expose him to 150 times more radiation than the airport scanner he passed through before boarding the plane. (Cosmic radiation gives a higher dose in the upper atmosphere than at ground level.) Incidentally, walking through the scanner delivers as much radiation as a tenth of a banana.

Nuclear power over the last 50 years, even including all accidents, has added only about a thousandth to our natural dose of background radiation – about as much as eating 20 bananas each. Even if you built your house right at the perimeter fence of your local nuclear plant, you would still receive many times less additional radiation than an airline crew member would consider entirely normal, for example. Coal-fired power stations in fact release far more radiation into the environment than nuclear power stations, due to trace radionuclides being concentrated into coal ash and blown away in dust and smoke.

This is why dozens of scientific studies have failed to find any association between leukaemia incidence and nuclear power stations. In the couple of occasions where an association has been found, this can only be a statistical fluke or the result of some unidentified causal factor other than radiation. We know this because radiation varies naturally from place to place to a far greater extent than levels are elevated near nuclear power stations, and yet there is no correlation with cancer incidence. (The only exception is a possible correlation between naturally occurring radioactive radon gas and lung cancer in smokers.)

This should not be at all surprising, because radiation is a relatively weak carcinogen – certainly when compared with self-inflicted evils such as smoking. Take the data from the survivors of the atomic bombings of Hiroshima and Nagasaki, all of whom were within 10 kilometres (6.2 miles) of the hypocentre of the atomic blasts on those fateful days in early August 1945. Their radiation doses were reconstructed carefully and their health for decades afterwards compared with control groups selected from nearby unaffected Japanese cities. You might assume that anyone witnessing the detonation of an atomic bomb at close quarters – assuming they survived the initial blast, fires and radiation sickness – would be pretty much doomed to die from cancer. Actually, this was not the case at all. The Hiroshima and Nagasaki atomic bomb survivors had about a 0.5 per cent increased risk of dying from cancer between 1950 and the year 2000. Of the entire studied group of 86,611, there were 10,127 cancer deaths by that time, of which 9,647 would have been expected anyway.[17]

This figure represents only an average, however. What is important about the A-bomb survivor data, and the reason why it remains the foundation for all radiological protection guidelines today, is that there is a clear statistical correlation between dose and cancer incidence above a certain threshold. Of the 625 people in the group who got the highest doses of above 2,000mSv, for example, 114 died of cancer by 2000 – double the expected baseline number of 56. Those in lower-dose groups had lower risks, however, while those receiving doses below 100mSv had no observable increase in risk at all. Out of the 68,467 people in the below-100mSv category, 7,657 died of cancer before 2000, compared with an expected number of cancer deaths of 7,655. The difference is too tiny to have any statistical meaning.

This latter conclusion is critically important. No convincing evidence has ever been obtained, despite many hundreds of studies, showing a statistically significant correlation between cancer incidence and radiation exposures of less than 100mSv. And yet, for the purposes of precaution, regulatory authorities and scientific agencies around the world assume that there is a linear association between exposure and risk even below the 100mSv threshold and indeed right down to zero. This 'linear no-threshold' (LNT) hypothesis is the subject of much scientific controversy, and even key UN agencies differ over its use. For example, UNSCEAR recently warned against the use of LNT for exactly the kind of study the WHO just published about Fukushima, described earlier in this chapter: all the subjects whose increased cancer risks are projected by the WHO report received doses well below the 100mSv level, below which no statistically significant correlations with cancer have actually been observed.

I don't propose to try to resolve the tortured LNT debate here, but it is important to appreciate that it essentially revolves around an argument over whether low-dose radiation has no risks at all, or has risks that are real but nevertheless too small to measure reliably. The important point for laypeople to remember is that adherents to either view would agree that the general alarm over small increases in radiation dose is scientifically unjustified. Recent studies have even suggested that low-dose radiation might have healing properties: a purported effect known as 'hormesis'. If this is true, those submerging themselves in Ramsar's hot spring radiation health spas and burying their arthritic bodies in Guarapari's radioactive black sand might be experiencing real rather than just imagined health benefits.

Chernobyl

Even after the worst nuclear power plant disaster of all time – the explosion and fire at Chernobyl in the then-USSR in April 1986 – the effects of radiation exposure were much less severe than everyone initially assumed. The oft-photographed deformed babies, for example, were to be found in equal numbers in areas contaminated by Chernobyl and those that received no fallout at all.[18] The total proven Chernobyl death toll currently stands at around 50, including the 28 emergency workers who died of acute radiation sickness at the time, and about 15 fatal thyroid cancers – in total about 6,000 children in Russia, Belarus and Ukraine suffered thyroid cancers that were successfully treated.

UNSCEAR's most important and least well-known conclusion is that, apart from the impacts mentioned above, there is "no persuasive evidence of any other health effect in the general population that can be attributed to radiation exposure".[19] That means that those Chernobyl 'victims' who wait in terror for the morning they wake to find a tumour sprouting on their body have had their lives blighted unnecessarily. All the scientific authorities now agree that the worst impact of Chernobyl has been social and psychological, due to fear of radiation and the dislocation effects of the exclusion zone, rather than the actual physical effects of radiation itself. Many more will die from social impacts such as suicides and alcoholism than will ever die from radiation. Anti-nuclear campaigners worsen this trauma by exaggerating the threats of radiation far beyond what is scientifically credible, doing real harm to people by labelling them as 'victims' and insisting that they are doomed to suffer and die early.

Unfortunately this lesson from Chernobyl does not appear to have been learned for Fukushima. What the affected communities desperately need is reliable information. The best science will tell them that they need not live in fear, that they should try to live normal lives as best they can when they are able to return to their homes, and that there will likely never be an observable increase in cancers as a result of the accident. The best thing local people could do, then, for themselves and their children in response to the Fukushima disaster, is simply to forget about it. It is not thoughtless or insensitive to say this; unnecessary fear is by far the greatest danger that people affected by Fukushima now face.

Deaths per terawatt-hour

There is a clear consensus in the scientific literature that, in terms of 'deaths per terawatt-hour', other conventional energy sources are more dangerous, by orders of magnitude, than is nuclear power. Oxides of sulphur and nitrogen released during coal combustion cause pulmonary and respiratory diseases, while particulates such as PM2.5 and PM10 from oil and coal are associated with congestive heart failure and cancers as well as bronchitis and asthma. Coal plants also release mercury and other toxins that can accumulate in the food chain. All told, according to the latest *Global Burden of Disease* study, ambient air pollution worldwide kills 3.2 million people per year.[20] The only serious debate in numbers terms is whether the death toll from coal is five hundred times worse than that from nuclear, or many thousands of times worse.

In a landmark 2013 paper, climatologists James Hansen and Pushker Kharecha calculated that the use of nuclear power between 1971 and 2009 avoided the premature deaths of 1.84 million people thanks to its air pollution benefits.[21] Ironically, 117,000 of those lives saved were in Germany, thanks to the clean operation of its 17 reactors (commissioned between 1975 and 1989) – and yet the Green Party rose to prominence during that same period primarily to demand the total elimination of nuclear power. Looking to the future, Hansen and Kharecha calculate that nuclear might prevent another 7 million deaths if deployed at large scale and in order to replace coal.

These figures ignore the potential lives saved by nuclear power because of its displacement of CO_2

emissions and consequent mitigation of climate change. Hansen and Kharecha do, however, tot up how much CO_2 has been saved: a total of about 64 billion tonnes over the last 35 years; equivalent to taking 430 large coal-fired power plants permanently off the grid. (This is about 8 per cent of humanity's total cumulative CO_2 emissions of 817 billion tonnes between 1971 and 2009.) Another 240 billion tonnes could be saved in future, they estimate, substantially increasing our ability to tackle global warming.

Looking at the air pollution mortality figures strongly suggests that it is untrue to say that nobody will die because of Fukushima. People will die, but not from radiation. Their lives will instead be shortened because of an increased reliance on fossil fuels due to post-Fukushima nuclear fear. This is nowhere more the case than in Japan, where after Fukushima every single reactor was closed down, and at the time of writing only two have restarted. Japan's 're-fossilization' has led to a massive surge in imports of natural gas and coal, and a consequent 50-million-tonne increase in annual CO_2 emissions.[22] The Japanese government is now proposing to abandon its climate change targets, acknowledging that meeting them will be impossible without nuclear power.[23]

Renewed nuclear fear has also led to nuclear exit plans in Belgium, Switzerland and the Netherlands, and even historically pro-nuclear France seems likely to reduce its proportion of nuclear-generated electricity from 80 per cent to 50 per cent thanks to a pre-election deal between President Hollande and the Greens. France currently has some of the lowest per-capita carbon emissions in Europe, about 4 tonnes lower than those in Germany (5.5 vs 9.3 tonnes[24]), but these will no doubt rise substantially if nuclear power generation is closed down as the French Green Party demands.

The German Experiment

The epicentre of the world's rejection of nuclear power surely lies in Germany. Germany's *Energiewende* nuclear-to-renewables energy transformation has become the emblem of a new global movement: environmentalists and renewables enthusiasts everywhere point to Germany as a model to be emulated worldwide.[25] The book *Clean Break: The story of Germany's energy transformation and what Americans can learn from it* has become an Amazon best-seller.

It all sounds terrifically inspiring, and some of it is: Germany's success in generating 5 per cent of its electrical power from solar photovoltaics in 2012, for example, was a truly impressive achievement. Wind and solar power combined generated 11.9 per cent of German electricity in 2012 (solar 4.6 per cent; wind 7.3 per cent). For comparison, nuclear generated nearly 18 per cent of Germany's power before Fukushima, but will now be eliminated entirely by 2022.[26]

If we do the implied maths, however, the picture is a little less rosy. Germany's 2012 production of solar power was 27 terawatt-hours, somewhat less than the 32TWh the country lost from shutting down eight of its nuclear plants after Fukushima.[27] It is therefore undeniable that Germany's emissions are higher now than they would have been had the new renewables investments been used to displace coal rather than nuclear power. Accordingly, Germany saw carbon emissions rise by 1.6 per cent in 2012, and is one of the only countries in Europe still building and opening new coal-fired power stations. In contrast, the more nuclear-friendly UK is committed to never building another unabated coal plant.

Indeed, Germany's flagship new lignite (brown coal) power station, which opened in mid-2012 at Neurath, is now the second-largest point source of CO_2 in the entire European Union.[28] The third-largest EU CO_2 source, Niederaussem, is also a coal-fired station in Germany. Although lignite is a particularly dirty form of coal, with higher emissions than the black coal anthracite, ministers and other VIPs at the opening ceremony praised the new coal plant's special new contribution to "climate protection" (in the words of the resulting press release),[29] as they joined hands to press a big green button and fire up the boilers. Germany is forecast to add another 4 gigawatts of coal to its grid in 2013,[30] and substantially more over the next decade.

I would join German environmentalists in celebrating the massive renewables scale-up the country has seen over the last few years: in that sense Germany has indeed helped light a pathway ahead. But the single and overriding priority for any environmentalist in the 400ppm world must be to quickly eliminate coal power. And this must surely take place first in rich countries such as Germany, as poorer countries in Asia and elsewhere still rely on cheap coal to fuel their much-needed economic growth. If Germany cannot lead the way in removing coal from its grid, then others must carry forward the torch of real environmental responsibility.

CHAPTER 5

Next generation: Nuclear 2.0

Like everyone I have ever met in the nuclear industry, I strongly believe that the next generation of reactors must be designed and engineered so that innocent people never again suffer the kind of accidents experienced at Chernobyl and Fukushima. Instead of granting life extensions for older plants, therefore, I think we should instead prioritize the deployment of new nuclear plants that are far safer than the previous generation. I call these Nuclear 2.0.

History lessons

Let's first examine some of the worst mistakes that were made in the past. The Soviet-era RBMK reactor involved in the Chernobyl accident was particularly ill-conceived. Designed to make bomb fuel as well as to provide steam to generate electricity, the reactor exhibited a 'positive void coefficient' – meaning, in simple terms, that the presence of boiling water in the reactor core (or a loss of water due to a leak) would increase its power generation in a deadly positive feedback. The Chernobyl reactor also had no proper containment, so once a steam explosion – followed a few seconds later by a hydrogen blast – had blown the lid off the reactor core and destroyed the flimsy outer building, the burning remains of the core were freely exposed to the atmosphere.

Chernobyl did not happen by accident, as it were: it was the direct result of a badly planned and disastrously executed safety experiment, during which the operators disabled the reactor's emergency cooling system before switching off electrical power to the pumps. They were trying to find out whether the coasting down of the steam turbines would provide enough electrical power to keep the coolant flowing for long enough to stabilize the reactor. The answer: clearly not.

One of nuclear history's little-known ironies is that just two weeks beforehand in the US a similar safety experiment on an entirely different type of reactor had resulted in a dramatically better outcome. Operators at Argonne National Laboratory's Experimental Breeder Reactor-II (EBR-II) also turned off the safety systems and shut down the coolant flow into their reactor, but in this case, as invited experts from around the world stood in the control room and watched with bated breath, the reactor simply shut itself down with no intervention.

The EBR-II experiment has passed into legend, not just because it was strikingly successful but also because the reactor programme of which it was a part was later cancelled by the Clinton Administration before its benefits could ever be realized. Anti-nuclear Democrats controlled Congress at the time, and, under the guidance of a powerful senator by the name of John Kerry, in 1994 all funding was cut to Argonne's Integral Fast Reactor (IFR) programme before a full-scale prototype could be built.

The reactor's key engineering characteristics are worth a closer look, however. First, the IFR is cooled by sodium, not water. Because sodium is a metal, in molten form it conducts heat 90 times more efficiently than water, and also boils at a much higher temperature (883°C / 1622°F). Unlike pressurized water reactors, therefore, the whole thing can operate at normal atmospheric pressure, simplifying both design and safety issues. Second, the sodium sits in a giant pool within which the reactor core is submerged, vastly reducing the likelihood of any coolant leak. (The pool is a single cast vessel with no welds or pipework penetrations.) Third, the reactor uses metal fuel, rather than the oxide fuel used in most reactors, further improving its passive safety. And fourth, it is a fast reactor, and therefore able to use fuel 100 times more efficiently than conventional reactors and even consume existing stockpiles of spent fuel and all the long-lived transuranics in nuclear waste (to which I return shortly).

There is a lot more to the technicalities than this, of course, and Dr Charles E. Till and Dr Yoon Il Chang, two of the IFR's key engineers, have produced a book called *Plentiful Energy: The story of the Integral Fast Reactor* – which is required reading for nuclear nerds like me. But the key take-away is that the IFR, as the EBR-II loss-of-coolant experiment showed, was designed for full passive safety. Even in the worst imaginable scenario, with a total loss of on-site power and an equally total failure of the core cooling system, the reactor would simply shut itself down. It simply could not melt down.

In a way, the IFR designers were learning from the anti-nuclear movement: they were trying to address its key concerns with clever engineering solutions. Given that people were worried about meltdowns, they designed a reactor that was passively safe. Also, since campaigners were concerned about proliferation and the possible diversion of plutonium for bomb production, a reprocessing system was designed which would make it technically and chemically extremely difficult to ever separate out bomb-grade fissile material, and where all recycled uranium and plutonium would be so radioactively hot that it could never be tampered with by any putative terrorist.

Proliferation is a real concern for many people, and rightly so. Nuclear is a classic example of a dual-use technology, but banning it altogether makes no more sense than trying to ban agriculturally vital nitrate fertilisers because they can also be used to manufacture terrorist car bombs. The challenge is for strong international regulation and transparency in the entire fuel cycle, and all signatories of the Non-Proliferation Treaty are expected to abide by these standards. Because Iran is refusing to do so, for example, it is clear that the country is indeed aiming for a nuclear weapon, and this certainly presents enormous challenges for the international community. The worst way of dealing with it, however, would be to ban nuclear power in the rest of the world in the hope that the Iranians would somehow thereby feel a moral pressure to give up their bomb programme. The challenge is to ban the military use of nuclear (the weapon), not the peaceful use (the power reactor). You have to beat swords into ploughshares, not try to ban them both.

The waste issue

Going back to the IFR, the third major concern of anti-nuclear activists that the engineers sought to address was the question of waste. It is important to establish at the outset that all nuclear waste generated by commercial nuclear plants in the US and other countries is today properly safeguarded and poses little danger to the environment.[1] You can see much of the US's nuclear waste safely immobilized in thick-walled dry cask storage behind barriers at the plant site. In France, all the nation's reprocessed high-level waste is currently stored under the floor of a single facility in La Hague about the size of a basketball court. You can walk around above it with no radiation shielding needed.

One of peoples' great fears about nuclear waste is the long half-lives of some isotopes. But if an element has a long half-life, it is not very radioactive, by definition. (A half-life is the period of time over which half of the mass of the element decays into something else.) That is why plutonium-239, which has a half-life of 24,000 years, can be held in a gloved hand and feels only slightly warm. If a radionuclide has a short half-life, like the 8-day half-life of iodine-131, it is very radioactive but quickly decays away. (I-131 was a big concern after both Fukushima and Chernobyl, but has virtually all now disappeared.) Other industrial processes, such as electronics, metals smelting or even manufacturing solar panels, produce toxic waste with an infinite half-life; it will remain dangerous to life until the end of the Universe. Radioactive waste is the only waste which becomes steadily safer over time, and yet oddly is the one people think we need to worry about for the longest. Such concerns do not seem to extend to common carcinogens and neurotoxins such as arsenic and mercury, which are also isotopically stable and therefore dangerous forever.

There is certainly a real issue about how best to deal with those fission products with medium-scale half-lives, primarily strontium-90 (half-life 29 years) and cesium-137 (half-life 30 years). These generate significant radiation during human-lifetime timescales. Cs-137 is now the radionuclide of most concern around Fukushima, because it emits strong gamma radiation during its decay and is volatile and biologically mobile – it can get into the body via drinking water and be absorbed by soft tissue.

Even Sr-90 and Cs-137 have their uses, however: the latter can be used in food irradiation, and both have applications in medical radiotherapy. Both also have multiple industrial uses – thousands of industrial applications of Cs-137 exist, from moisture-density gauges to well-logging devices used by the drilling industry to characterize sub-surface rock strata.[2] Even if we do decide to throw away these isotopes in deep repositories, they both disappear almost completely in a few centuries; not the millions of years that most people think.

Still, I accept that waste is a serious concern for many people, so the fact that fast reactors like the IFR can actually burn the longest-lived elements in nuclear waste should be a huge incentive to deploy them. These long-lived elements are the 'transuranics' such as americium, californium and neptunium, which are the reason why deep geological disposal facilities are designed to be stable for a million years. (Americium-241 has another important use: it may have already saved your life as a crucial component of your household smoke detector.) Fast reactors are also able to burn uranium-238, both directly via fission and indirectly via conversion to plutonium-239, which then fissions in turn. This means that not only can fast reactors – deployed at scale – 'solve' the nuclear waste 'problem', but that they can also run entire countries for centuries on uranium that has already been mined and for which there is little other use.

The numbers are in fact quite startling. When I looked at the uranium inventory of the UK I was surprised to find that the country already has enough nuclear fuel – if burned in fast reactors – to run the nation at full current electric output for 500 years. In the US, existing uranium stockpiles could run the country for more than a millennium.[3] Remember: that is without mining even another scrap of uranium, and if it ever runs short there's plenty more out there yet to be discovered. That is why the IFR designers' book is called *Plentiful Energy* – even if all fossil fuel runs out (or we decide to leave it in the ground so as not to fry the planet), there is no conceivable shortage of nuclear fuel to burn in fast reactors.

And uranium is not the only potential nuclear fuel: we also have the option of using thorium. In recent years a dedicated group of enthusiasts has rediscovered the enormous potential for using thorium in 'liquid fluoride thorium reactors' (LFTRs, pronounced 'lifters'). Many designs have already been produced, and indeed back in the 1950s thorium was considered the ideal nuclear fuel for civilian purposes. As its name suggests, in a LFTR the fuel is not held in a solid core but is dissolved in circulating molten salts, meaning that waste fission products can be removed (and new fuel added) on an ongoing basis without the reactor being shut down.

LFTRs also have excellent safety features. My favourite is the use of a 'plug' that would melt if the molten mass got too hot for any reason, draining it away into a protected lower tank that would stop any fissioning and cool the whole lot down. It's a clever idea: the plug is a frozen wedge of salt in a pipe at the bottom of the core tank, cooled by an external fan. If power is lost for some reason and might threaten to overheat the LFTR, the fan stops, the plug melts, and the salts all drain away. The fuel can't melt down for the simple reason that it is already molten. No China Syndrome here.

There are enormous quantities of thorium sitting around looking for a good use. Thorium is a by-product of rare-earth mining, and is currently considered a nuisance. One tonne of refined thorium – which you might imagine as a sphere about the size of one of those big gym balls people use for exercise training – would generate a gigawatt of electrical energy for a year: enough to power a large city, and at a trivial fuel cost of around £195,000 ($300,000), according to recent estimates. Current proven reserves total about 3 million tonnes, while as much as 120 trillion tonnes may exist in the Earth's crust, enough to run civilization for far longer than it is likely to last.

Even the more conventional light water reactors are today evolving fast. So-called 'third generation plus' (Gen III+) reactors such as the French EPR, the Westinghouse AP1000 and GE Hitachi's ESBWR all have improved safety designs, such as the ESBWR's gravity-driven core cooling system – large tanks high up inside the containment building would provide enough cooling water to keep the core from overheating even in a total-loss-of-external-power scenario of the kind that occurred at Fukushima. The standard for all Gen III+ reactors is passive safety for 72 hours at least, with no operator intervention needed, thanks to natural circulation systems which can remove heat continually from the core. That way they could survive a Fukushima-type scenario intact with no release of radiation.

Too expensive?

All this leaves just one major piece out of the puzzle: cost. The two European Pressurized Reactors (EPRs) proposed for the UK's Hinkley site are reported to have a capital cost of £14 billion (about $21 billion).[4] That is real money by anyone's reckoning, and the need to build new nuclear reactors in the UK to replace older plants is so economically daunting that it has required the redesign of the whole electricity market so that nuclear generators (and other capital-intensive low-carbon generators) can be persuaded to invest. The omens are not good: two existing EPR projects, at Flamanville in France and Olkiluoto in Finland, are both years over time and 2 billion euros over budget.

Part of the reason why nuclear projects have always been so costly is that each one is a gigantic, complex construction project, and even the slightest delay quickly becomes prohibitively expensive. The solution to this may be to move to modular designs in which the components of nuclear power plants are produced and assembled in factories rather than on site, allowing economies of scale and uniformly high manufacturing standards. Modularity is easier if reactors are smaller: the pressure vessels for standard 1-gigawatt pressurized water reactors are now so massive that they can generally only be transported by ship.

GE Hitachi (GEH) has designed its PRISM (Power Reactor Innovative Small Modular) reactor – the modern redesign of the original Integral Fast Reactor that was cancelled by Clinton and Kerry back in 1994 – to be fully modular. Each reactor module would produce about 300 megawatts of electric power, making it a quarter of the size of the lumbering EPR, and could be brought to site by road on the back of a lorry. Construction time is estimated to be 36 months. Several modules could be knitted together in one site, analogous to the way that additional turbines can be added to a wind farm. GEH is now proposing that six modules be brought together into an 'advanced recycling centre' for dealing with nuclear waste while generating a hefty 1.8 gigawatts of electric power.

Today there is a whole host of 'small modular reactor' (SMR) designs competing for the attention of any electric utilities interested in affordable and reliable clean power. In early 2013 the US Department of Energy announced $150 million of support for Babcock & Wilcox's mPower SMR design: modules of this 180-megawatt miniature pressurized water reactor will be fully assembled in a factory and delivered complete to Tennessee Valley Authority's Clinch River site, where they will be housed – with others added as necessary – in underground bunkers and will become operational as early as 2021.

Other SMRs are even smaller: NuScale's Power Module is rated at 45MW, conceived for grids that cannot handle the massive gigawatt-scale electrical output of today's giant reactors. Needless to say, all are designed for full passive safety. Westinghouse's SMR (essentially a smaller version of its AP1000, scaled down to 225MW) is designed to be so safe that "no operator intervention is required for seven days" after a theoretical worst-case-scenario accident.[5]

The SMR revolution, and all the other competing Gen III+ and Gen IV designs, show that nuclear technology has not stood still. It is as much a mistake, therefore, to judge the potential for new nuclear on the basis of accidents such as Fukushima and Chernobyl as it is to judge the safety of the new Airbus A380 on the crash record of the 1970s-era McDonnell Douglas DC-10. There is no doubt that new designs available today are dramatically safer than those of the past, and that the potential for severe accidents in future is vastly reduced.

Cost may also be less of a concern in non-Western countries where large-scale engineering is more familiar. In China, EPRs are being built at Taishan, and, unlike their relatives in France and Finland, the new reactors are being built to time and on budget. Moreover, their quoted capital cost, about £6.8 billion ($10.4 billion), is about half that proposed for the UK reactors, even though the rated power output is the same.[6] Cheaper labour may have something to do with it, as may the fact that all large construction projects – bridges or high-speed rail, for instance – can be completed in China at much lower cost than in the West. China's nuclear regulation is as strict as elsewhere, so savings are not being made by cutting corners on safety.

Although reactors can be built in Asia for half the price of those in the West, cost, in my view, is the one issue that still clearly remains to be solved. The nuclear industry now has an enormous challenge to bring costs down, just as solar and other renewables manufacturers have been able to do – or it will find itself priced out of increasingly competitive low-carbon energy markets. Modularity may be part of the answer: we must eventually be able to churn out reactors like sausages worldwide if we are to deploy them at the scale necessary. It is time for us to get ambitious.

CHAPTER 6

The spectre of climate change

What happens if we fail, and the nuclear renaissance runs into the sand – either because of a renewed anti-nuclear movement or because new reactors are too costly to be worth bothering with, or because of some other factor? This is where we must run some final numbers.

The coal-based scenario

First we need a baseline – I will use the US Energy Information Administration's projections to 2030. The EIA projects a 250-per-cent increase in wind power and a 400-per-cent increase in solar power by that year: a major scale-up but still not enough to prevent global CO_2 emissions in 2030 rising to 40.6 billion tonnes, about 28 per cent higher than today. This EIA baseline projection also includes nuclear accounting for 14 per cent of global electricity in 2030, a proportion essentially unchanged from today.[1]

But let's suppose that all nuclear power plants are shut down one by one between now and 2030, and are replaced by coal. (Coal substitutes one-for-one for nuclear because it is also an always-on, baseload-supplying electricity generator.) This would mean an additional 4.4 billion tonnes of annual CO_2 emissions, which would otherwise have been avoided by 2030's operating nuclear plants. Adding this to the EIA's baseline projection gives us a total CO_2 emission of 45 billion tonnes in 2030[2] – a hefty 42 per cent above today's level.[3]

How much does this matter for the climate? To work this out we have to run the emissions numbers through a climate model. Like most of us, I don't have one at home, so here I benefitted from the kind assistance of Jason Lowe, one of the UK's top climate modellers, based at the British Meteorological Office's Hadley Centre. (Lowe's work is repeatedly cited in all the major scientific climate assessments, and he has been a contributing author to the Intergovernmental Panel on Climate Change [IPCC]'s working group II.) For the emissions scenario inputs, Lowe wrote an algorithm which smoothed out the increase to give us the 2030 figure, kept emissions rising until an assumed 2040 peak, and then took emissions down again by 3 per cent per year until 2100.[4] The temperature outputs came from a climate model called MAGICC, which has a climate sensitivity of about 3°C (5.4°F), so is in the mid-range of most IPCC climate models.[5]

So if the nuclear naysayers are right, what kind of a world are we heading towards? According to Lowe's climate model, this is a hotter world that likely misses the internationally recognized 2°C (3.6°F) target; there is a 50:50 chance of about 2.6°C (4.7°F) rise above pre-industrial temperatures by 2100, and a 1-in-10 chance of a hefty 3.6°C (6.5°F) by the same date in this scenario. Remember, this is an optimistic projection: I have arbitrarily peaked emissions in 2040 and made them fall dramatically thereafter, but if CO_2 continues to rise throughout the century in a world that abandons nuclear power, we could be looking at five or six degrees Celsius of warming.

Given that I wrote a book which devotes a separate chapter to assessing the impacts of each degree of global warming, up to (as the title has it) 6 degrees, I can quickly précis what this nuclear-free world would look like. First, the 50:50 relatively lucky outcome – my chapters about the world for 2 to 3 degrees Celsius (3.6 to 5.4 degrees Fahrenheit) include the following impacts:

- Extreme dryness in the American west, possibly including a reappearance of ancient deserts in Nebraska and neighbouring states.
- Total disappearance of the Arctic's permanent sea-ice cap, and ecological disaster for ice-dependent animal species such as the polar bear.
- Major degradation of coral reefs due to more frequent bleaching; extinction threatened for a third of animals and plants worldwide.
- Stronger hurricanes striking across wider areas of the global coastline.

- Accelerating sea-level rise, threatening major coastal cities and dooming low-lying island states.
- Spreading desertification and drought belts throughout the sub-tropics, affecting in particular southern Europe, southern Africa and northern China.
- Possible collapse of rainforest ecosystems due to severe drying, heat stress and fire.
- Severe water depletion in areas dependent on snowmelt, such as the US south-west, or on glacial runoff, such as Peru and Pakistan.

I'd better stop there. You get the picture. If we are unlucky, and the nuclear-free world bequeaths a global warming outcome of 3 to 4 degrees Celsius (5.4 to 7.2 degrees Fahrenheit), we can add in the following:

- Global collapse in food production due to drought and heat stress of crops in main breadbasket areas.
- Total disappearance of glacial ice in the Andes and Alps, and over substantial portions of the Himalayas; elimination of snow from many mountain regions.
- Possible release of vast quantities of additional carbon dioxide (and methane, which is far worse) from thawing Siberian permafrost, potentially including the melting of sub-sea methane hydrates as oceans heat up.
- Rise in ocean acidification, which is already occurring due to added uptake of carbon dioxide.

A change in pH (acidity) means trouble for shelled marine life, including coccolithophores, which make a notable percentage of the Earth's oxygen.
- Drought/flood impacts of severe weather globally, due to intensification of the hydrological cycle, much heavier rainfall events and longer droughts.

And, no doubt, much more. Anyone familiar with the evolving scientific literature on climate impacts – both from climate modelling outputs and paleo-climate investigations – will be able to add to this list.

But perhaps I'm being unfair. Doubtless anti-nuclear campaigners would argue that the EIA's projections for renewables deployments are unduly pessimistic, and we should install vastly more wind and solar plants in order to offset the otherwise inevitable rise in fossil-fuel consumption that would result from phasing out nuclear power. To represent this perspective as honestly as possible, for the next climate modelling exercise I will use figures produced by Greenpeace and the Global Wind Energy Council (GWEC), an industry body.

The renewables scale-up scenario

Greenpeace and the GWEC have produced energy scenarios for 2030 that are vastly more ambitious than those of the US Energy Information Administration. Greenpeace/GWEC project wind power generation increasing by 1,500 per cent above today's levels, and solar growing by 9,500 per cent.[6] Combined, this is more than 10 times the renewables scale-up projected by the EIA for 2030.[7] These are enormous renewables investments, which surely must be at the absolute upper end of what can be considered realistic. By 2030, wind power would be generating 22 per cent of global electricity, and solar 17 per cent.

The scale of the task can perhaps best be represented in area terms rather than abstract numbers. In the Greenpeace/GWEC scenario, by 2030 wind farms would cover, according to my calculations, about 1 million square kilometres (386,000 square miles) – 17 per cent of that offshore – of the globe.[8] That is equivalent to the combined areas of the US states of Pennsylvania, Ohio, Virginia, Tennessee, Kentucky, Indiana, Maine, South Carolina, West Virginia, Maryland, Hawaii, Massachusetts, Vermont, New Hampshire, New Jersey, Connecticut, Delaware and Rhode Island. Alternatively, it is also about as much as Texas and New Mexico combined, or twice the area of Spain.[9] Solar power plants would cover another 50,000 or so square kilometres (19,305 square miles);[10] there would be 2,500 concentrating solar power plants of the scale of California's Ivanpah site in the world's hot deserts, with the rest of the area devoted to solar PV.[11]

Whether or not this scale of renewables ramp-up is technically or politically do-able is not the issue; the point of this exercise is to calculate whether in the context of what Greenpeace and GWEC are suggesting we still need nuclear power in order to avoid dangerous climate change. Nor will I consider other potential problems with a renewables-only energy system, such as the obvious challenges of intermittency and the consequent need for backup power or large-scale electricity storage. Let's assume for the sake of argument that these problems have technical solutions, which are implemented across the board by 2030. Greenpeace does actually provide cost estimates for the suggested deployment, which add up to a global investment total of about £5.7 trillion ($8.8 trillion) between 2011 and 2030.[12] I am happy to take all these figures at face value.

Now let's look at the carbon picture. 'Scenario Greenpeace' would displace about 6.5 billion tonnes of additional CO_2 per year by 2030 compared with the EIA reference scenario, thanks to the gigantic renewables programme. This, however, falls to only 1.9 billion tonnes displaced when nuclear's projected CO_2 benefits are removed from the equation.[13] According to my calculations, therefore, Greenpeace's scenario has emissions rising by 22 per cent by 2030 rather than the EIA's 28 per cent; hardly a great improvement. This projection resembles a global version of Germany's Energiewende: a massive renewables investment that could have been reducing emissions makes little difference because wind and solar displace nuclear instead of fossil fuels.[14]

Needless to say, this scenario does not save the world from rapid climate change either, even if the warming is a little less rapid than in the first baseline example (the 'coal-based scenario'), which was also nuclear-free but with a lower investment in renewables. If we again model an idealized emissions pathway, which sees CO_2 rise (to 39 billion tonnes) by 2030, and afterwards peak and begin to decline a decade later, the result is a 50:50 chance of a climate warming of 2.4°C (4.3°F), and a 1-in-10 chance of a warming of 3.3°C (5.9°F). Again we miss the 2-degree-Celsius targets that Greenpeace and other environmental groups are campaigning for, and to which the world's governments are committed.

The conclusion is clear: if nuclear is removed from the picture, even the greatest imaginable investment in renewables reduces eventual global warming by at best a couple of tenths of a degree Celsius when compared with business as usual. This is why, as I have repeatedly argued, those environmentalists who insist on a renewables-only prescription to tackle global warming are gambling at very poor odds with our planetary biosphere. I think we can and must do better than this.

CHAPTER 7

All of the above

One of the most inspiring initiatives I have been involved with recently was an under-the-radar effort to combine the forces of the UK's low-carbon industries in supporting the British government's proposed energy bill. The result was, to my knowledge, a global first: a joint press release by RenewableUK (the wind and marine energy trade association), CCSA (the Carbon Capture and Storage Association) and the Nuclear Industry Association, issued on 5 November 2012.[1]

As *The Independent* reported in a page 2 splash: "The leaders of Britain's nuclear, wind and tidal industries today put aside years of mutual suspicion and antipathy with an unprecedented joint appeal to ministers not to abandon their commitment to combat climate change." The story continued: "The letter marks the first time that Britain's nuclear and renewables industries have made common cause together. Significantly, the joint approach has won the backing of the environmental group Greenpeace – despite its long-standing opposition to nuclear power. It welcomed the 'unity' demonstrated by the low- and zero-carbon industries who signed the letter and their goal to take carbon almost completely out of the electricity system by 2030."[2]

I wrote an accompanying editorial in the same newspaper. "One unintended benefit of this joint approach by the three trade associations will perhaps be to wrong-foot antis of all political stripes – the Greens who oppose nuclear, the increasingly vocal anti-wind lobby and those sceptics who insist CCS will never be viable and is not worth supporting. Make no mistake: opposing low-carbon technologies is an implicit vote for a high-carbon energy system, and opponents must recognize this real-world trade-off."[3]

And here is the reaction from Greenpeace UK executive director John Sauven: "This letter shows that whilst different industries will have differing preferences for the exact mix of energy technologies, there is unity from across huge swathes of the business community on the need for a clear goal in the energy bill to take carbon almost completely out of the electricity system by 2030. As well as helping reduce the risks of climate change, this would help counter the overwhelming sense of confusion currently hanging over the direction of UK energy policy."

This unprecedented collaboration between the wind and nuclear industries helped give the UK government the support it needed: at its third reading on 5 June 2013 the Energy Bill was passed by the House of Commons by an overwhelming vote of 396 to 8. As Ed Davey, Secretary of State for Energy and Climate

Change, said in response: "We're already bound by law to cut emissions across the whole UK economy by 50 per cent by 2025, and the Energy Bill will bring about substantial decarbonization of the power sector as part of that." This was a clear 'all of the above' strategy: "Long-term contracts for low carbon will give renewables, nuclear and CCS the chance to compete against conventional power stations, and will be backed by a tripling in support for clean energy technologies by 2020."[4]

A low-carbon global vision

We might argue about the specifics, but overall I think the UK has got the right idea. So let's dare to dream a little. What might happen if we extended this 'all of the above' strategy to a global scale? What would low-carbon 'unity' (to borrow John Sauven's word) be able to achieve worldwide if those backing nuclear power and those promoting renewables joined forces instead of always fighting each other and thereby sealing our dependence on fossil fuels? This is our final scenario, one where everyone drops his or her technology tribalism – whether anti-wind, anti-solar, anti-nuclear or anti-anything – and low-carbon industries are promoted worldwide in a coordinated and ambitious way.

In order to make the calculation, I will once again adopt Greenpeace and the Global Wind Energy Council's mega-ambitious figures for renewables deployments, as outlined in the last chapter. But this time I will also add in a massive nuclear scale-up in tandem with the major investments in wind and solar power. The nuclear industry's own figures are too unambitious for this purpose, so I will use my own – I suggest we aim for a global reactor fleet of about 1,000 nuclear plants by 2030. Today we have 420 operating reactors, so – to allow for some decommissioning of old plants – this implies about four new reactors opened each month somewhere in the world between now and then.[5] Note that this scenario still includes much greater investment in renewables than nuclear: wind and solar would generate about 12,000 terawatt-hours per year of electricity in 2030, while nuclear would contribute just over 8,000TWh.

Is this realistic? It certainly implies a major effort by the nuclear industry to globally streamline and modularize the supply chain in order to get costs down and produce the necessary number of finished reactors. It also requires a move by the world's regulators to speed up their work: new, safer reactor designs will need to be brought online quickly, without years or even decades of regulatory delay. We have some historical precedent for this build rate: between 1980 and 2000, France commissioned over 60GW of nuclear power, about three 1-gigawatt plants per year – and that was just in one European country. China already plans capacity increases that would take its fleet up to 200GW by 2030,[6] and I would roughly propose a doubling of this target, with a similar level of ambition for India – a country which already plans to have 60 reactors running by 2030 and as many as 500 operational by 2050.[7]

Crunch the resulting numbers and in my 'all of the above' scenario we get a global electricity supply that is 82 per cent carbon-free by 2030.[8] That really would be an extraordinary achievement. Consequently, instead of rising continuously for at least the next two-and-a-half decades, global emissions would peak in around 2021 and then begin to decline. If we assume the same rate of decline thereafter as in the other two scenarios outlined in the last chapter, by 2030 we will have delivered a worldwide 3-per-cent cut on 2011 emissions.[9] According to our climate model, the world then at last has a 50:50 chance of hitting the magic 2-degrees-Celsius target (the unlucky 10-per-cent outcome is 2.8°C [5°F]).

Here is what was agreed at the Cancún climate change UN summit in 2010, in which I participated as a delegate for the Maldives:[10]

> [The Conference of the Parties] recognizes that deep cuts in global greenhouse gas emissions are required according to science, and as documented in the Fourth Assessment Report of the Intergovernmental Panel on Climate Change, with a view to reducing global greenhouse gas emissions so as to hold the increase in global average temperature below 2°C above pre-industrial levels, and that Parties should take urgent action to meet this long-term goal, consistent with science and on the basis of equity...[11]

My conclusion is that only an 'all of the above' strategy can meet these goals. It is environmentally responsible because it holds the increase in global temperature at the 2 degrees Celsius maximum rise above pre-industrial levels. It is also 'equitable', as the Cancún agreement demands, because it allows for significant energy-use growth in developing countries, so that these nations can continue to win the battle against poverty.

That is why I called this book 'Nuclear 2.0' – because only with a new generation of safer nuclear reactors, added to an even greater push for renewables, can we keep the world from reaching the dangerous levels of global warming above 2 degrees. What I am of course advocating here is the oldest idea in the world: that with unity comes strength, and that when confronted with a common and immense global challenge like climate change we should put aside old enmities and join together to tackle it.

Dropping dogma

I recognize that it will not be easy for the environmental movement as a whole to drop its decades-long anti-nuclear position. People who have dedicated their entire lifetimes to campaigning against nuclear, and whole organizations that were founded in the 1960s- and 1970s-era anti-nuclear zeal, cannot realistically be expected to turn on a dime and adopt an opposite perspective. It will be a difficult and traumatic experience for the environmental movement to make this shift, just as it was for me to do so on a personal level.

But today the anti-nuclear movement is already a shadow of its former self. As I hope I have made clear in this book, the advent of global warming has changed the environmentalist landscape utterly. Neither Friends of the Earth nor Greenpeace any longer have dedicated anti-nuclear staff in the UK, even though policies have not formally changed. This is because both organizations are canny enough to understand that the greatest environmental challenge of our time is fossil fuels, and both have instead reassigned their formidable expertise and resources into campaigning against oil and coal.

The same transition has taken place internationally. The Sierra Club, whose website contains barely a mention of nuclear power, has mounted a fantastically successful 'Beyond Coal' campaign, partly as a result of which 150 proposed coal-fired power plants in the US have been defeated and will now never be built. The Club has celebrated the retirement of 147 US coal plants in recent years, and aims to get 375 more – a third of the total – retired by 2020. The 350.org climate campaign, which was inspired by the pro-nuclear climatologist James Hansen, has successfully rallied thousands of protesters against the proposed Keystone pipeline, which would take dirty oil from the Canadian tar sands to refineries in the US. In his landmark climate change speech delivered at Georgetown University on 25 June 2013, President

Obama strongly backed an 'all of the above' strategy that included renewables, nuclear, energy efficiency and natural gas as a 'transition fuel'. This is a welcome 'war on coal' in all but name.

It is said that hope springs eternal; but not today, for we do not have much time. My 'all of the above' scenario projects how the world might be in 2030, only 17 years from now. This is not the world of our distant descendants; for most of us it will be our own world still, which we together will bring into being. I hope this book has shown that during this period we can indeed continue to make dramatic progress in eradicating poverty, and that carbon emissions can at the same time stabilize and begin to fall, as is necessary to protect the climate. Combine these two imperatives, and future generations can surely enjoy as many lifetime opportunities as I want for my own children.

Notes*

Chapter 1 • How we got to where we are

1 Figures from International Energy Agency (IEA), *CO_2 Emissions from Fuel Combustion – Highlights, 2012 Edition*, Excel file, from: www.iea.org/publications/freepublications/publication/name,32870,en.html
2 IEA (2010), *Energy Poverty: How to make modern energy access universal.* Special early excerpt of the World Energy Outlook 2010 for the UN General Assembly on the Millennium Development Goals, page 11. www.worldenergyoutlook.org/media/weowebsite/2010/weo2010_poverty.pdf
3 Oliver August, 'A hopeful continent'. *The Economist*, 2 March 2013. www.economist.com/news/special-report/21572377-african-lives-have-already-greatly-improved-over-past-decade-says-oliver-august
4 Fareed Zakaria (2011), *The Post American World: And the rise of the rest.* Penguin: London.
5 Fareed Zakaria (2011), *The Post American World: And the rise of the rest.* Penguin, London.
6 Keisuke Sadamori / IEA. 'Medium-term Coal Market Report 2012'. Presentation on 18 December 2012. www.iea.org/newsroomandevents/speeches/121218MCMR2012_presentation_KSK.pdf
7 Figures from *BP Statistical Review of World Energy June 2012.*
8 IEA (2011), *Climate & Electricity Annual Data and Analyses 2011*, Figure 6, page 83. www.iea.org/publications/freepublications/publication/Climate_Electricity_Annual2011.pdf

* These notes are available as a pdf with live hyperlinks at http://uit.co.uk/files/b-nuclear/links.htm

9 Calculated using 'net summer capacity' via US Energy Information Administration (EIA) 'Independent Statistics & Analysis' / Electricity / Electricity generating capacity: www.eia.gov/electricity/capacity
10 UK coal-fired generating capacity is currently 23GW – see Table 5.7, Chapter 5 (page 140) of *Electricity: Digest of United Kingdom Energy Statistics (DUKES)*, Department of Energy & Climate Change. www.gov.uk/government/uploads/system/uploads/attachment_data/file/65818/DUKES_2013_Chapter_5.pdf
11 IEA, 'Medium-Term Coal Market Report 2012 Factsheet', 18 December 2012. www.iea.org/newsroomandevents/news/2012/december/name,34467,en.html
12 All figures in this section are from *BP Statistical Review of World Energy June 2013.*
13 EIA 'Independent Statistics & Analysis' / International Energy Statistics / Petroleum: www.eia.gov/cfapps/ipdbproject/iedindex3.cfm?tid=50&pid=53&aid=1&cid=&syid=2013&eyid=2013&freq=M&unit=TBPD
14 'Supply shock from North American oil rippling through global markets', IEA press release, 14 May 2013: www.iea.org/newsroomandevents/pressreleases/2013/may/name,38080,en.html
15 For an example, see Figure 1, page 3, in Indur Goklany (2012) 'Humanity Unbound: How fossil fuels saved humanity from nature and nature from humanity'. *Policy Analysis*, 20 December. www.cato.org/doc-download/sites/cato.org/files/pubs/pdf/pa715.pdf

Chapter 2 • The carbon challenge

1 Michael Shellenberger, Ted Nordhaus and Jesse Jenkins, 'Energy Emergence: Rebound and backfire as emergent phenomena'. *The Breakthrough*, 17 February

2011: http://thebreakthrough.org/archive/new_report_how_efficiency_can

2 Chris Goodall and Mark Lynas, 'It's a myth that wind turbines don't reduce carbon emissions'. *The Guardian* Environment blog, 26 September 2012: www.guardian.co.uk/environment/blog/2012/sep/26/myth-wind-turbines-carbon-emissions

3 Energy figures in this chapter are from *BP Statistical Review of World Energy June 2013* except where stated.

4 This suggests a CO_2 intensity of energy displaced of 600g CO_2/kWh, which seems reasonable since wind is more likely to substitute for fluctuating gas than baseload coal. Source: Greenpeace / Global Wind Energy Council (2012), *Global Wind Energy Outlook 2012*, pages 18 and 19. www.gwec.net/wp-content/uploads/2012/11/GWEO_2012_lowRes.pdf

5 Vaclav Smil (2010), *Energy Myths and Realities: Bringing Science to the Energy Policy Debate.* AEI Press: Washington, DC.

6 'The Plowboy Interview with Amory Lovins', *Mother Earth*, November/December 1977. www.motherearthnews.com/Renewable-Energy/1977-11-01/Amory-Lovins.aspx?page=14#ixzz2NcynCUkI

7 Roger Harrabin, 'Renewable energy: Burning US trees in UK power stations'. BBC News Science & Environment, 28 May 2013: www.bbc.co.uk/news/science-environment-22630815

8 Roger Harrabin, 'Biofuels: MPs to consider subsidies for power stations'. BBC News Science & Environment, 6 March 2013: www.bbc.co.uk/news/science-environment-21672840

Chapter 3 • The N-word

1 Mark Lynas, 'Nuclear power: a convert'. *New Statesman*, 30 May 2005. www.newstatesman.com/node/150738
2 Barry Brook, 'Golf balls and elephants – energy density in 9 seconds', posting on BraveNewClimate.com, 22 June 2011: http://bravenewclimate.com/2011/06/22/golf-balls-elephants-energy-density
3 'The Austrian people said "no" to nuclear energy'. TheNuclearPowerPlant.net, 2010: www.nuclear-power-plant.net/index.php?lang=en&item=history
4 'Austrian nuclear plant goes 100% solar – at 0.003% capacity'. DepletedCranium.com, 21 August 2009: http://depletedcranium.com/austrian-nuclear-plant-goes-100-solar-at-03-capacity
5 'That day in December: The story of nuclear prohibition in Australia'. DecarboniseSA.com, blog posting on 12 September 2012: http://decarbonisesa.com/2012/09/12/that-day-in-december-the-story-of-nuclear-prohibition-in-australia
6 'List of cancelled nuclear plants in the United States'. Wikipedia webpage: http://en.wikipedia.org/wiki/List_of_canceled_nuclear_plants_in_the_United_States
7 This currently stands at about 310GW. See EIA, 'Independent Statistics & Analysis' / Total Energy / Annual Energy Review, Table 8.11c (Electric Net Summer Capacity: Electric Power Sector by Plant Type, 1989-2011): www.eia.gov/totalenergy/data/annual/index.cfm#electricity
8 EIA, 'Independent Statistics & Analysis'/ Total Energy / Annual Energy Review, Table 7.3 (Coal Consumption by Sector, 1949-2011): www.eia.gov/totalenergy/data/annual/index.cfm
9 EIA statistics are not disaggregated into coal and other fossil fuels until 1989. See EIA, 'Independent

Statistics & Analysis' / Total Energy / Annual Energy Review, Table 8.11a Excel data (Electric Net Summer Capacity: Total [All Sectors], 1949-2011): www.eia.gov/totalenergy/data/annual/index.cfm

10 Increase from 771TWh/year to 1341TWh/yr. See EIA, 'Independent Statistics & Analysis' / Total Energy / Annual Energy Review, Table 8.2b Excel data (Electricity Net Generation: Electric Power Sector, 1949-2011): www.eia.gov/totalenergy/data/annual/index.cfm

11 See EIA, 'Independent Statistics & Analysis' / Total Energy / Annual Energy Review, Table 8.2b Excel data (Electricity Net Generation: Electric Power Sector, 1949-2011): www.eia.gov/totalenergy/data/annual/index.cfm. This gives a total solar generation of 5.248GWh in 1984; total electricity generation was 2,416TWh in that year.

Chapter 4 • The case against: nuclear accidents and radiation

1 This part of the account relies heavily on the excellent ebook *Fukushima: the First Five Days* by Leslie Corrice.

2 Rick Wallace, 'Fukishima boss Masao Yoshida breaks silence on disaster'. *The Australian*, 11 August 2012. www.theaustralian.com.au/news/world/fukushima-boss-masao-yoshida-breaks-silence-on-disaster/story-fnb1brze-1226448211757

3 Readers may have heard that since the accident Yoshida received treatment for oesophagal cancer and sadly died on 9 July 2013. Given the latency period of radiation-induced cancers of 10 years or so, it is extremely unlikely that this is linked to his Fukushima experience. Instead it seems probable that Yoshida's cancer was smoking-related, given his quoted accounts

of handing out 'smokes' during the crisis.

4 Sandia National Laboratories (2012), *Fukushima Daiichi Accident Study* (status as of April 2012). http://energy.sandia.gov/wp/wp-content/gallery/uploads/Fukushima_SAND2012-6173.pdf
5 See 'Fukushima Accident 2011', World Nuclear Association: www.world-nuclear.org/info/fukushima_accident_inf129.html for core damage figures. Sandia/ORNL (see note 4 above) says 1.5m- (5ft-) deep intrusion into the concrete base from modelling.
6 From TEPCO press release attachment: 'Exposure dose distribution', Table 2: www.tepco.co.jp/en/press/corp-com/release/betu12_e/images/120731e0401.pdf
7 Average dose from The National Diet of Japan (2012), *The Official Report of the Fukushima Nuclear Accident Independent Investigation Commission (NAIIC): Executive Summary*, Chapter 4, Table 4.1-3. http://warp.da.ndl.go.jp/info:ndljp/pid/3856371/naiic.go.jp/wp-content/uploads/2012/08/NAIIC_Eng_Chapter4_web.pdf
8 United Nations Scientific Committee on the Effects of Atomic Radiation (UNSCEAR) (2012), *Report of 59th Session*, 21-25 May 2012.
9 The National Diet of Japan (2012), *The Official Report of the Fukushima Nuclear Accident Independent Investigation Commission (NAIIC): Executive Summary*, Chapter 4, Table 4.1-3. http://warp.da.ndl.go.jp/info:ndljp/pid/3856371/naiic.go.jp/wp-content/uploads/2012/08/NAIIC_Eng_Chapter4_web.pdf
10 'No immediate health risks from Fukushima nuclear accident says UN expert science panel'. UNSCEAR press release, 31 May 2013. www.unis.unvienna.org/unis/en/pressrels/2013/unisinf475.html
11 'Global report on Fukushima nuclear accident details health risks'. World Health Organization (WHO) news

release, 28 February 2013. www.who.int/mediacentre/news/releases/2013/fukushima_report_20130228/en

12 S. M. Javad Mortazavi, 'High background radiation areas of Ramsar, Iran': www.ecolo.org/documents/documents_in_english/ramsar-natural-radioactivity/ramsar.html

13 P. Andrew Karam (2002), 'The high background radiation area in Ramsar, Iran'. Proceedings of the Waste Management 2002 Symposium. www.wmsym.org/archives/2002/proceedings/10/434.pdf

14 This is shown in the film *Pandora's Promise* as microsieverts per hour, which is the real-time reading given by radiation dosimeters. I have transposed this to millisieverts per year by multiplying by 8760 and then dividing by 1000.

15 Nathaniel Rich, 'Showdown at the Airport Body Scanner'. *The New York Times*, 25 May 2013: http://opinionator.blogs.nytimes.com/2013/05/25/showdown-at-the-airport-body-scanner

16 'Airport Screening', Health Physics Society factsheet, May 2011. http://hps.org/documents/airport_screening_fact_sheet.pdf

17 Dale L. Preston et al. (2004), 'Effect of recent changes in atomic bomb survivor dosimetry on cancer mortality risk estimates'. *BioOne* 162 (4), 377-89. See Table 3. www.bioone.org/doi/full/10.1667/RR3232

18 World Health Organization (2006), *Health Effects of the Chernobyl Accident and Special Health Care Programmes*: Report of the UN Chernobyl Forum Expert Group 'Health'. http://whqlibdoc.who.int/publications/2006/9241594179_eng.pdf

19 UNSCEAR (2008), *Sources and effects of ionizing radiation*, Volume II. www.unscear.org/docs/reports/2008/11-80076_Report_2008_Annex_D.pdf

20 Stephen S. Lim et al. (2012), 'A comparative risk

assessment of burden of disease and injury attributable to 67 risk factors and risk factor clusters in 21 regions, 1990-2010: A systematic analysis for the Global Burden of Disease Study 2010'. *The Lancet* 380 (9859), 2224-60. See Table 3. www.thelancet.com/journals/lancet/article/PIIS0140-6736(12)61766-8/fulltext

21 Pushker A. Kharecha and James E. Hansen (2013), 'Prevented mortality and greenhouse gas emissions from historical and projected nuclear power'. *Environmental Science & Technology*, DOI:10.1021/es3051197. http://decarbonisesa.files.wordpress.com/2013/04/prevented-mortality-and-greenhouse-gas-emissions.pdf

22 National Institute for Environmental Studies, Japan. *Japan's National Greenhouse Gas Emissions in Fiscal Year 2011 (Preliminary Figures): Executive Summary.* www.nies.go.jp/whatsnew/2012/20121205/pdf/gaiyou-e.pdf

23 'Abe looking to renege on emissions pledges'. *The Japan Times*, 25 January 2013. www.japantimes.co.jp/news/2013/01/25/national/abe-looking-to-renege-on-emissions-pledge/#.UQ-LuqX7VO3

24 IEA, *CO_2 Emissions from Fuel Combustion, 1971-2010*, Excel file, from: www.iea.org/publications/freepublications/publication/name,4010,en.html

25 For example, see this from the World Resources Institute (WRI): '5 achievements from Germany's "Energiewende"', Lutz Weischer, 13 May 2013: http://insights.wri.org/news/2013/05/5-achievements-germanys-energiewende

26 See my blog: 'Germany's "Energiewende" – the story so far', 15 January 2013, for numbers and German-language original sources: www.marklynas.org/2013/01/germanys-energiewende-the-story-so-far

27 The 32TWh figure comes from the *BP Statistical Review of World Energy June 2012* nuclear spreadsheet.

28 Vera Eckert and Christoph Steitz, 'Germany's clean energy drive fails to curb dirty brown coal', 29 April 2013. *Planet Ark*: http://planetark.org/wen/68524

29 'Efficient and highly flexible: BoA 2&3 makes important contribution to transforming German energy industry and climate protection'. RWE Power AG press release, 15 August 2012. www.rwe.com/web/cms/en/2320/rwe-power-ag/press-releases/press-release/?pmid=4008220

30 Stefan Nicola, 'Germany to add most coal-fired plants in two decades, IWR says'. Bloomberg.com, 27 February 2013: www.bloomberg.com/news/2013-02-27/germany-to-add-most-coal-fired-plants-in-two-decades-iwr-says.html

Chapter 5 • Next generation: Nuclear 2.0

1 Although no country has yet built and begun operating a deep geological repository for civilian nuclear waste (the US has a successful military repository in operation in New Mexico, the Waste Isolation Pilot Plant or WIPP), the idea that existing waste presents a serious danger to the environment is overblown. Waste is either being cooled in pools or in safe dry cask storage, although different countries have very different approaches. For a summary of the current situation, see the World Nuclear Association webpage 'Radioactive Waste Management': www.world-nuclear.org/info/Nuclear-Fuel-Cycle/Nuclear-Wastes/Radioactive-Waste-Management

2 US Environmental Protection Agency, 'Radiation Protection' / Cesium: www.epa.gov/radiation/radionuclides/cesium.html

3 This figure assumes that depleted uranium from fuel enrichment, uranium-238 and other actinides in nuclear fuel can all be recycled in fast reactors – U-238, for example, is transmuted via neutron capture into Pu-239 and then fissioned. The '500 years' figure was picked up by *The Guardian* (see Duncan Clark, 'New generation of nuclear reactors could consume radioactive waste as fuel', *The Guardian*, 2 February 2012: www.theguardian.com/environment/2012/feb/02/nuclear-reactors-consume-radioactive-waste) and confirmed by David MacKay, chief scientist at the UK Department of Energy and Climate Change.
4 Guy Chazan, 'EDF raises heat at Hinkley C over price'. *Financial Times*, 13 March 2013. www.ft.com/cms/s/0/45610f04-8afa-11e2-b1a4-00144feabdc0.html#axzz2T5wLeL8H
5 'Westinghouse SMR Features', WestinghouseNuclear.com: www.westinghousenuclear.com/smr/features.htm
6 'First fuel produced for Chinese EPR'. *World Nuclear News*, 11 March 2013. www.world-nuclear-news.org/ENF-First_fuel_produced_for_Chinese_EPR-1103134.html

Chapter 6 • The spectre of climate change

1 EIA (2011), *International Energy Outlook 2011*. www.eia.gov/forecasts/archive/ieo11
2 Assuming nuclear baseload substitutes for coal baseload globally, and coal has a carbon intensity of 958g/CO_2 per kWh. The latter figure is a world average given by the IEA for 2010 – see IEA, *CO_2 Emissions from Fuel Combustion – Highlights*, Summary table 'CO_2 emissions per KWh from electricity generation using coal/peat', page 114, which can be downloaded from: www.iea.org/publications/freepublications/publication/name,32870,en.html. The EIA reference

scenario projects nuclear at 4545TWh/yr in 2030, resulting in CO_2 abatement of 4.355 billion tonnes/yr, and projects global CO_2 at 40.6 bn. tonnes/yr; 2011 emissions were 31.639 bn. tonnes. Adding back nuclear's avoided CO_2 gives a total 2030 emission of 44,995 million tonnes.

3 'Today's level' means 2011 CO_2 figures, the last real figures in the EIA 2011 spreadsheets; 2012 and onwards are projections. A new *International Energy Outlook* was published in late July 2013, too late to be included here. See www.eia.gov/forecasts/ieo

4 This is intended to be a conservative estimate of emissions. There is no actual reason why emissions should peak in 2040 in this scenario and drop thereafter – but the principle here is that post-2040 energy use and emissions paths are too uncertain to be able to forecast accurately, hence this rather idealized scenario. However, with no peak in emissions, temperature rise would of course be much higher. The key figure inputting into the climate model is cumulative emissions during the entire century, so emissions paths post-2030 are crucial.

5 The MAGICC model parameters are as per Lowe et al. (2009), 'How difficult is it to recover from dangerous levels of global warming?', *Environmental Research Letters* 4 (1), DOI:10.1088/1748-9326/4/1/014012. http://nora.nerc.ac.uk/6700/1/lowe_et_al_09.pdf. Spreadsheet of emissions scenario inputs and temperature outputs for this whole exercise available from: www.marklynas.org/wp-content/uploads/2013/06/avoid_basesheet_210613.xlsx

6 See tables on page 295 of Greenpeace International / Global Wind Energy Council (GWEC) / European Renewable Energy Council (EREC) (2012): *Energy [R]evolution: A sustainable world energy outlook.*

www.greenpeace.org/international/Global/international/publications/climate/2012/Energy%20Revolution%202012/ER2012.pdf

7 Wind 2030 projection is from the Greenpeace/GWEC report *Global Wind Energy Outlook 2012* (page 11) (www.gwec.net/wp-content/uploads/2012/11/GWEO_2012_lowRes.pdf), and solar projections are from Greenpeace / GWEC / EREC: *Energy [R]evolution: A sustainable world energy outlook* (tables on page 295), as in note 6 of this chapter.

8 Assuming 2011 conversion rates of 1.98TWh electricity produced per GW installed of wind. Greenpeace, in *Global Wind Energy Outlook 2012*, projects 6289TWh of wind in 2030 in addition to today's generation, implying an additional installed capacity above 2011 by 2030 of 3171GW. Paul Denholm et al. (2009), *Land-use Requirements of Modern Wind Power Plants in the United States*, National Renewable Energy Laboratory technical report (page 22) (www.nrel.gov/docs/fy09osti/45834.pdf), averages out total area of 160 US wind farms per unit of capacity and ends up with a figure of 0.34 square km (0.13 square miles) per MW of nameplate capacity. (Remember: animal grazing, agriculture, etc. can continue in between the turbines, so this is not land used exclusively for wind.) Therefore, with 3171GW installed this would equal a land take of 1,078 million square km (416 million square miles). Bigger turbines in future will not much change this equation, as they need to be spaced commensurately further apart.

9 A good list of US states and land area in square km is here: http://geography.about.com/od/usmaps/a/states-area.htm. Build your own list!

10 For PV see US Department of Energy, 'Energy Efficiency and Renewable Energy: PV FAQs' factsheet

(www.nrel.gov/docs/fy04osti/35097.pdf), which gives 5-25 square km (1.9-9.6 square miles) per TWh/yr of generation. The Greenpeace solar PV 2030 projection is for 2,634TWh/yr.

11 Nevada Solar One produces 0.134TWh/yr over 400 acres. Ivanpah is scheduled to produce 1.079TWh/yr over 3,500 acres. The efficiencies are roughly comparable despite the different technologies employed (solar mirrors vs. parabolic troughs). See 'Concentrating Solar Power Projects', National Renewable Energy Laboratory (www.nrel.gov/csp/solarpaces/project_detail.cfm/projectID=20) for this and other CSP details. In UAE, Shams 1 will produce 0.21TWh/yr over 617 acres, a similar rate of conversion. The Greenpeace 2030 projection for CSP is 2,672TWh/yr. For solar PV the Greenpeace 2030 projection is 2634TWh/yr.

12 See Greenpeace / GWEC / EREC: *Energy [R]evolution: A sustainable world energy outlook*, Table 12.15, page 296. www.greenpeace.org/international/Global/international/publications/climate/2012/Energy%20Revolution%202012/ER2012.pdf

13 Assuming a CO_2 displacement of 606g/CO_2 per kWh, which is reasonable as wind and solar do not displace coal baseload so much as gas, and gas has half the carbon intensity of coal. The figure I use here comes from the Greenpeace/GWEC report *Global Wind Energy Outlook 2012* (www.gwec.net/wp-content/uploads/2012/11/GWEO_2012_lowRes.pdf), which projects avoided emissions in 2030 of 4.007 billion tonnes with 6,678TWh/yr of wind, so 1TWh = 0.600 million tonnes CO_2 displaced. The total deployment is 10,463TWh/year in addition to the wind and solar generation that is already included in the EIA's reference scenario for 2030, and for which CO_2 reductions are

already included in the EIA baseline. I assume solar and wind have the same CO_2 displacement factor. The result is a 38.7 bn. tonnes annual CO_2 emission in 2030: 22% higher than 2011 emissions. Nuclear emissions abatement removed (added to derive the total) is 4.355 bn. tonnes, as per first nuclear-free scenario described here.

14 Greenpeace / GWEC / EREC's *Energy [R]evolution: A sustainable world energy outlook* claims CO_2 tonnage savings of 19.2 bn tonnes by 2030. But these are predicated on highly questionable assumptions, including vastly reduced use of cars globally, reduction in aviation, global deployment of passivhaus constructions, wide uptake of heat pumps and so on, which reduce energy use overall as compared with the reference scenario. It claims savings from power generation specifically of 9.3 billion tonnes of CO_2 by 2030, although how these are derived is not quantified. Either way, for any given scenario, if you remove the mitigation benefits of nuclear the emissions will be worse – that is my point.

Chapter 7 • All of the above

1 'Joint letter from heads of RenewableUK, CCS association and Nuclear Industry Association on need for 2030 carbon goal'. *The Guardian*, 5 November 2012: www.guardian.co.uk/environment/interactive/2012/nov/05/letter-renewableuk-ccs-nuclear-carbon

2 Oliver Wright, 'Nuclear, wind and wave power chiefs in joint appeal on green energy'. *The Independent*, 5 November 2012. www.independent.co.uk/news/uk/politics/nuclear-wind-and-wave-power-chiefs-in-joint-appeal-on-green-energy-8281122.html

3 To be clear, I still think CCS is worth supporting with public funds. I have not included it in these scenarios

because full-scale CCS has yet to be deployed anywhere, so there is no realistic basis on which to make quantified assumptions about potential CO_2 reductions. I have also ignored shale gas and its role in displacing coal, as I assume this to be taking place equally in all scenarios.

4 'Energy Bill completes Commons passage with overwhelming majority'. Department of Energy & Climate Change press release, 5 June 2013. www.gov.uk/government/news/energy-bill-completes-commons-passage-with-overwhelming-majority

5 In 2009 the 420-odd reactors comprised 377GW installed capacity. Assuming 200 decommissioned are decommissioned by 2030, about 800 have to be built to reach the required 1,000 total by then. With 17 years left, 47 reactors have to be built per year, or about 4 per month. I'm also assuming 1,200GWe per reactor for a 1,000-strong reactor fleet.

6 'Nuclear Power in China', World Nuclear Association: www.world-nuclear.org/info/Country-Profiles/Countries-A-F/China--Nuclear-Power

7 'Nuclear Power in India', World Nuclear Association: www.world-nuclear.org/info/Country-Profiles/Countries-G-N/India

8 The EIA projects a baseline of 12,136TWh of zero-carbon electricity from nuclear, hydro and renewables in 2030. I have added 10,463TWh from Greenpeace's wind and solar projections, and 3,707TWh for my nuclear, both additional to what is already included in the EIA's reference scenario. EIA projects total electricity consumption of 31,940TWh in 2030, and this 'all of the above' scenario has 26,307TWh provided by nuclear, hydro and renewables, which is 82% of the total. Nuclear figures are based on an assumption of 6.878TWh/yr of electricity per

GW of nuclear. This ratio is based on 2009 figures, pre-Fukushima to avoid low capacity factors of politically motivated closures. 2009 had 377GW installed nuclear capacity, production of 2594TWh.

9 For wind and solar CO_2 displacement factors, see note 13, Chapter 6. Wind and solar CO_2 displaced is 6278 million tonnes/yr and there is also 3551 m. tonnes/yr avoided from nuclear. (These are for nuclear and renewables generation additional to the EIA reference scenario, to avoid double-counting.) EIA has 2030 emissions at 40,640 m. tonnes/yr, this 'all of the above' scenario has 30,809 m. tonnes/yr, a 3% cut on 2011 total of 31,639 m. tonnes/yr. Yay.

10 I was Climate Advisor to President Mohamed Nasheed of the Maldives from 2009 to 2011.

11 United Nations Framework Convention on Climate Change (2011), *Report of the Conference of the Parties on its sixteenth session, held in Cancun from 29 November to 10 December 2010.* http://unfccc.int/resource/docs/2010/cop16/eng/07a01.pdf#page=2

Index

Also published by UIT

ISBNs:
9780954452933 (paperback)
9781906860011 (hardback)
384 pages

ISBNs:
9781906860059 (paperback)
9781906860073 (hardback)
384 pages

About UIT

News:
Forthcoming titles, events, reviews, interviews, podcasts, etc.
www.uit.co.uk/news

Join our mailing lists:
Get email newsletters on topics of interest.
www.uit.co.uk/subscribe

How to order:
Get details of stockists and online bookstores. If you are a bookstore, find out about our distributors or contact us to discuss your requirements.
www.uit.co.uk/order

Send us a book proposal:
If you want to write – even if you have just the kernel of an idea at present – we'd love to hear from you. We pride ourselves on supporting our authors and making the process of book-writing as satisfying and as easy as possible.
www.uit.co.uk/for-authors